FOREIGN POLICY OF THE UNITED STATES

U.S. NUCLEAR COOPERATION WITH OTHER COUNTRIES

FOREIGN POLICY OF THE UNITED STATES

Additional books in this series can be found on Nova's website
under the Series tab.

Additional E-books in this series can be found on Nova's website
under the E-books tab.

FOREIGN POLICY OF THE UNITED STATES

U.S. NUCLEAR COOPERATION WITH OTHER COUNTRIES

GABRIEL A. OLIVEIRA
EDITOR

Nova Science Publishers, Inc.
New York

For permission to use material from this book please contact us:
Telephone 631-231-7269; Fax 631-231-8175
Web Site: http://www.novapublishers.com

Library of Congress Cataloging-in-Publication Data

U.S. nuclear cooperation with other countries / editor, Gabriel A. Oliveira.
p. cm.
Includes index.
ISBN 978-1-61209-910-1 (hardcover)
1. Nuclear energy--Government policy--United States. 2. Nuclear energy--International cooperation. I. Oliveira, Gabriel A. II. Title: United States nuclear cooperation with other countries.
HD9698.U52.U58 2011
333.792'40973--dc22

2011004963

Published by Nova Science Publishers, Inc. † New York

CONTENTS

PREFACE

Under existing law (Atomic Energy Act of 1954), all significant U.S. nuclear cooperation with other countries requires a peaceful nuclear cooperation agreement. Significant nuclear cooperation includes the transfer of U.S.-origin special nuclear material subject to licensing for commercial, medical and industrial purposes. Such agreements, which are "congressional-executive agreements" requiring congressional approval, do not guarantee that cooperation will take place or that nuclear material will be transferred, but rather set the terms of reference and authorize cooperation. This new book examines U.S. nuclear cooperation with other countries including Australia, Russia, India, South Korea and the United Arab Emirates.

Chapter 1- In order for the United States to engage in civilian nuclear cooperation with other states, it must conclude a framework agreement that meets specific requirements under section 123 of the Atomic Energy Act (AEA). The AEA also provides for exemptions to these requirements, export control licensing procedures, and criteria for terminating cooperation. Congressional review is required for section 123 agreements; the AEA establishes special parliamentary procedures by which Congress may act on a proposed agreement.

Chapter 2- The United States has 26 agreements in force for peaceful nuclear cooperation. Under the U.S. Atomic Energy Act of 1954, as amended, these agreements are a prerequisite to certain aspects of U.S. nuclear cooperation with other cooperating partners. GAO was asked to (1) quantify the amount and value of U.S. nuclear exports facilitated by these agreements, (2) assess U.S. efforts to support the U.S. nuclear industry's ability to compete for sales, and (3) examine U.S. nuclear industry challenges to exporting. To conduct this work, GAO reviewed and assessed data collection efforts by U.S. agencies from 1994 through 2008, analyzed available data, and interviewed U.S. industry representatives and U.S. and foreign government officials.

Chapter 3- Australia and the United States have cooperated in the peaceful use of nuclear energy since the mid-1950s. The framework for this cooperation is a civilian nuclear cooperation agreement as required by section 123 of the Atomic Energy Act. President Obama transmitted the proposed text of the latest renewal agreement to Congress on May 5, 2010, along with the required Nuclear Proliferation Assessment Statement (NPAS) and his determination that the agreement promotes U.S. national security. Congress has 30 days of continuous session for consultations with the Administration, followed by an additional 60 days of continuous session to review the agreement. If not opposed by a joint resolution of disapproval or other legislation, then the agreement will be considered approved at the end of this time period. Congress also has the option of adopting either a joint resolution of approval with (or without) conditions or standalone legislation that could approve or disapprove the agreement. On November 30, 2010, the House passed H.R. 6411 by voice vote. The bill would approve the agreement even if the required congressional review period is not reached. The Senate has not yet acted on its version of the bill (S. 3844).

The United States and Australia first concluded a civilian nuclear cooperation agreement in 1957. That agreement was updated in 1979. Australia sells around 36% of its $1 billion in uranium exports to the United States. The United States is also a major processor of Australian uranium sold to other countries. Australia does not currently possess any nuclear power plants, but it operates one research reactor.

Chapter 4- The United States and Russia signed a civilian nuclear cooperation agreement on May 6, 2008. President Bush submitted the agreement to Congress on May 13. The agreement was withdrawn from congressional consideration by President George W. Bush on September 8, 2008, in response to Russia's military actions in Georgia. President Obama transmitted the proposed text of the agreement to Congress on May 10, 2010, along with the required Nuclear Proliferation Assessment (NPAS) and his determination that the agreement promotes U.S. national security. Congress has 30 days of continuous session for consultations with the Administration, followed by an additional 60 days of continuous session to review the agreement. If not opposed by a joint resolution of disapproval or other legislation, then the agreement will be considered approved at the end of this time period.

Chapter 5- India, which has not signed the Nuclear Nonproliferation Treaty (NPT) and does not have International Atomic Energy Agency safeguards on all of its nuclear material, exploded a "peaceful" nuclear device in 1974, convincing the world of the need for greater restrictions on nuclear trade. The United States created the Nuclear Suppliers Group (NSG) as a direct response to India's test, halted nuclear exports to India a few years later, and worked to convince other states to do the same. India tested nuclear weapons again in 1998. However, President Bush announced July 18, 2005, he would "work to achieve full civil nuclear energy cooperation with India" and would "also seek agreement from Congress to adjust U.S. laws and policies," in the context of a broader partnership with India.

Chapter 6- A South Korean consortium recently signed a contract to provide four commercial nuclear reactors to the United Arab Emirates (UAE), signaling a new role for South Korea in the world nuclear energy market. The $20 billion deal indicates that South Korea has completed the transition from passive purchaser of turn-key nuclear plants in the 1 970s to major nuclear technology supplier, capable of competing with the largest and most experienced nuclear technology companies in the world. The South Korean government reportedly has established a goal for South Korea to capture 20% of the world nuclear power plant market during the next 20 years, and the importance placed by Seoul on the UAE contract was underscored by South Korean President Lee Myung-bak's presence at the signing ceremony in the UAE.

Chapter 7- The United Arab Emirates (UAE) has embarked on a program to build civilian nuclear power plants and is seeking cooperation and technical assistance from the United States and others. During 2008 and early 2009, the Bush Administration and the UAE government negotiated and signed a memorandum of understanding and a proposed bilateral agreement on peaceful nuclear cooperation pursuant to Section 123 of the Atomic Energy Act (AEA) of 1954. Then-U.S. Secretary of State Condoleezza Rice signed the proposed agreement on peaceful nuclear cooperation with the UAE January 15, 2009. Deputy Secretary of State James Steinberg signed a new version of the agreement May 21; the Obama Administration submitted the proposed agreement to Congress the same day.

In: U.S. Nuclear Cooperation with Other Countries
Editor: Gabriel A. Oliveira
ISBN: 978-1-61209-910-1

Chapter 1

NUCLEAR COOPERATION WITH OTHER COUNTRIES: A PRIMER

Paul K. Kerr and Mary Beth Nikitin

SUMMARY

In order for the United States to engage in civilian nuclear cooperation with other states, it must conclude a framework agreement that meets specific requirements under section 123 of the Atomic Energy Act (AEA). The AEA also provides for exemptions to these requirements, export control licensing procedures, and criteria for terminating cooperation. Congressional review is required for section 123 agreements; the AEA establishes special parliamentary procedures by which Congress may act on a proposed agreement.

WHAT IS A "SECTION 123" AGREEMENT?

Under existing law (Atomic Energy Act [AEA] of 1954; P.L. 95-242; 42 U.S.C. §2153 et seq.)[1] all significant U.S. nuclear cooperation with other countries requires a peaceful nuclear cooperation agreement.[2] Significant nuclear cooperation includes the transfer of U.S.-origin special nuclear material[3] subject to licensing for commercial, medical, and industrial purposes. Such agreements, which are "congressional-executive agreements" requiring congressional approval, do not guarantee that cooperation will take place or that nuclear material will be transferred, but rather set the terms of reference and authorize cooperation. The AEA includes requirements for an agreement's content, conditions for the President to exempt an agreement from those requirements, presidential determinations and other supporting information to be submitted to Congress, conditions affecting the implementation of an agreement once it takes effect, and procedures for Congress to consider and approve the agreement.

Section 123 of the AEA requires that any agreement for nuclear cooperation meet nine nonproliferation criteria and that the President submit any such agreement to the House Committee on Foreign Affairs and the Senate Committee on Foreign Relations. The Department of State is required to provide the President an unclassified Nuclear Proliferation Assessment Statement (NPAS), which the President is to submit, along with the agreement, to those two committees. The State Department is also required to provide a classified annex to the NPAS, prepared in consultation with the Director of National Intelligence. The NPAS is meant to explain how the agreement meets the AEA nonproliferation requirements. The President must also make a written determination "that the performance of the proposed agreement will promote and will not constitute an unreasonable risk to, the common defense and security."

Requirements Under the Atomic Energy Act

Section 123 of the AEA specifies the necessary steps for engaging in nuclear cooperation with another country:

- **Section 123a.** states that the proposed agreement is to include the terms, conditions, duration, nature, and scope of cooperation and lists nine criteria that the agreement must meet. It also contains provisions for the President to exempt an agreement from any of several criteria described in that section and includes details on the kinds of information the executive branch must provide to Congress.
- **Section 123b.** specifies the process for submitting the text of the agreement to Congress.
- **Section 123c.** specifies how Congress approves cooperation agreements that are limited in scope (e.g., do not transfer nuclear material or cover reactors larger than 5 MWe). This report does not discuss such agreements.
- **Section 123d.** specifies how Congress approves agreements that do cover significant nuclear cooperation (transfer of nuclear material or reactors larger than 5 MWe), including exempted agreements.

Section 123a., paragraphs (1) through (9), lists nine criteria that an agreement must meet unless the President determines an exemption is necessary. These include guarantees that

- safeguards on transferred nuclear material and equipment continue in perpetuity;
- full-scope International Atomic Energy Agency (IAEA) safeguards are applied in non-nuclear weapon states;
- nothing transferred is used for any nuclear explosive device or for any other military purpose; except in the case of cooperation agreements with nuclear weapon states, the United States has the right to demand the return of transferred nuclear materials and equipment, as well as any special nuclear material produced through their use, if the cooperating state detonates a nuclear explosive device or terminates or abrogates an IAEA safeguards agreement;
- there is no retransfer of material or classified data without U.S. consent;

- physical security on nuclear material is maintained;
- there is no enrichment or reprocessing by the recipient state of transferred nuclear material or nuclear material produced with materials or facilities transferred pursuant to the agreement without prior approval;
- storage for transferred plutonium and highly enriched uranium is approved in advance by the United States; and
- any material or facility produced or constructed through use of special nuclear technology transferred under the cooperation agreement is subject to all of the above requirements.

Exempted vs. Non-exempted Agreements

The President may exempt an agreement for cooperation from any of the requirements in Section 123a. if he determines that the requirement would be "seriously prejudicial to the achievement of U.S. non-proliferation objectives or otherwise jeopardize the common defense and security." The AEA provides different requirements, conditions, and procedures for exempt and non-exempt agreements.[4] To date, all of the section 123 agreements in force are non-exempt agreements.[5] Prior to the adoption of P.L. 109-401, the Henry J. Hyde United States-India Peaceful Atomic Energy Cooperation Act of 2006, the President would have needed to exempt the nuclear cooperation agreement with India from some requirements of Section 123 a. P.L. 109-401, however, exempted nuclear cooperation with India from some of the AEA's requirements.[6]

CONGRESSIONAL REVIEW

Under the AEA, Congress has the opportunity to review a 123 agreement for two time periods totaling 90 days of continuous session.[7] The President must submit the text of the proposed nuclear cooperation agreement, along with required supporting documents (including the unclassified NPAS) to the House Foreign Affairs Committee and the Senate Foreign Relations Committee. The President is to consult with the committees "for a period of not less than 30 days of continuous session." After this period of consultation, the President is to submit the agreement to Congress, along with the classified annex to the NPAS and a statement of his approval of the agreement and determination that it will not damage the national security interests of the United States. This action begins the second period, of 60 days of continuous session. In practice, the President has sent the agreement to Congress at the beginning of the full 90-day period, which begins on the date of transmittal. Typically, the 60-day period has immediately followed the expiration of the 30-day period. The President transmits the text of the proposed agreement along with a letter of support with a national security determination, the unclassified NPAS, its classified annex, and letters of support for the agreement from the Secretary of State and the Nuclear Regulatory Commission.

If the President has not exempted the agreement from any requirements of section 123 a., it becomes effective at the end of the 60-day period unless, during that time, Congress adopts a joint resolution disapproving the agreement and the resolution becomes law. If the agreement is an exempted agreement, Congress must adopt a joint resolution of approval and it must become law by the end of the 60-day period or the agreement will not enter into force. At the beginning of this 60-day period, joint resolutions of approval or disapproval, as appropriate, are to be automatically introduced in each house. During this period, the committees are to hold hearings on the proposed agreement and "submit a report to their respective bodies recommending whether it should be approved or disapproved." If either committee has not reported the requisite joint resolution of approval or disapproval by the end of 45 days, it is automatically discharged from further consideration of the measure. After the joint resolution is reported or discharged, Congress is to consider it under expedited procedures, as established by section 1 30.i. of the AEA.

Section 202 of P.L. 110-369, the United States-India Nuclear Cooperation Approval and Nonproliferation Enhancement Act, which President Bush signed into law October 8, 2008, amended section 123 of the AEA to require the President to keep the Senate Foreign Relations Committee and the House Foreign Affairs Committee "fully and currently informed of any initiative or negotiations relating to a new or amended agreement for peaceful nuclear cooperation."

EXPORT LICENSING

The AEA sets out procedures for licensing exports to states with whom the United States has nuclear cooperation agreements. (Sections 126, 127, and 128 codified as amended at 42 U.S.C. 2155, 2156, 2157.) Each export of nuclear material, equipment, or technology requires a specific export license or other authorization. The Nuclear Regulatory Commission (NRC) is required to meet criteria in sections 127 and 128 in authorizing export licenses. These criteria are as follows:

- Application of IAEA safeguards to any material or facilities proposed to be exported, material or facilities previously exported, and to any special nuclear material used in or produced through the use thereof (these are not full-scope safeguards, but safeguards required under Article III.2 of the nuclear Nonproliferation Treaty [NPT]).
- Nothing exported can be used for any nuclear explosive device or for research on or development of any nuclear explosive device.
- Recipient states must have adequate physical security on "such material or facilities proposed to be exported and to any special nuclear material used in or produced through the use thereof."
- Recipient states are not to retransfer exported nuclear materials, facilities, sensitive nuclear technology, or "special nuclear material produced through the use of such material" without prior U.S. approval.
- Recipient states may not reprocess or alter in form or content exported nuclear material or special nuclear material produced though the use of exported nuclear material without prior U.S. approval.

- The foregoing conditions must be applied to any nuclear material or equipment that is produced or constructed under the jurisdiction of the recipient by or through the use of any exported sensitive nuclear technology.
- Section 128 requires that recipient non-nuclear-weapon states must have full- scope IAEA safeguards.

The President must judge that the proposed export or exemption will "not be inimical to the common defense and security" or that any export of that type "would not be inimical to the common defense and security because it lacks significance for nuclear explosive purposes." The executive branch may also consider other factors, such as "whether the license or exemption will materially advance the nonproliferation policy of the United States by encouraging the recipient nation to adhere" to the NPT; whether "failure to issue the license or grant the exemption would otherwise be seriously prejudicial" to U.S. nonproliferation objectives; and whether the recipient nation has agreed to conditions identical to those laid out in Section 127.

Section 126b. (2) contains a provision for the President to authorize an export in the event that the NRC deems that the export would not meet section 127 and 128 criteria. The President must determine "that failure to approve an export would be seriously prejudicial to the achievement of U.S. nonproliferation objectives or otherwise jeopardize the common defense and security." In that case, the President would submit his executive order, along with a detailed assessment and other documentation, to Congress for 60 days of continuous session. After 60 days of continuous session, the export would go through unless Congress were to adopt a concurrent resolution of disapproval.[8]

Section 128b.(2) contains a provision for the President to waive termination of exports by notifying the Congress that the state has adopted full-scope safeguards or that the state has made significant progress toward adopting such safeguards, or that U.S. foreign policy interests dictate reconsideration. Such a determination would become effective unless Congress were to adopt a concurrent resolution of disapproval within 60 days of continuous session.

Additionally, section 129b.(1) forbids the export of "nuclear materials and equipment or sensitive nuclear technology" to any country designated as a state sponsor of terrorism.[9] Section 129b.(3) allows the President to waive this provision.

TERMINATION OF COOPERATION

Section 1 29a. of the AEA requires that the United States end exports of nuclear materials and equipment or sensitive nuclear technology to any non-nuclear-weapon state that, after March 10, 1978, the President determines to have detonated a nuclear explosive device; terminated or abrogated IAEA safeguards; materially violated an IAEA safeguards agreement; or engaged in activities involving source or special nuclear material and having "direct significance" for the manufacture or acquisition of nuclear explosive devices, and "has failed to take steps which, in the President's judgment, represent sufficient progress toward terminating such activities."

Section 1 29a. also requires that the United States halt exports to any nation the President determines to have materially violated the terms of an agreement for cooperation with the U.S.; assisted, encouraged, or induced any non-nuclear weapon state to obtain nuclear explosives or the materials and technologies needed to manufacture them; or re-transferred or entered into an agreement for exporting reprocessing equipment, materials or technology to a non-nuclear weapon state, unless in connection with an international agreement to which the United States subscribes.

The President can waive termination of exports if he determines that "cessation of such exports would be seriously prejudicial to the achievement of United States nonproliferation objectives or otherwise jeopardize the common defense and security." The President must submit his determination to Congress, which is then referred to the House Committee on Foreign Affairs and the Senate Foreign Relations Committee for 60 days of continuous session. The determination becomes effective unless Congress adopts a concurrent resolution opposing the determination.

End Notes

[1] The Atomic Energy Act (AEA) was amended by the Nuclear Nonproliferation Act of 1978 (NNPA) (P.L. 95-242) to include stringent nonproliferation requirements for significant U.S. nuclear exports. For example, the act required nonnuclear-weapon states to have full-scope International Atomic Energy Agency safeguards as a condition for entering into nuclear cooperation agreements with the United States. For existing and future agreements, the NNPA added a provision for Congress to review export licenses. The act also included a provision for halting exports if a country tested a nuclear device, violated safeguards agreements, or continued nuclear weapons-related activities.

[2] Section 57 b. (2) of the AEA allows for limited forms of nuclear cooperation related to the "development or production of any special nuclear material outside of the United States" without a nuclear cooperation agreement if that activity has been authorized by the Secretary of Energy following a determination that it "will not be inimical to the interest of the United States." Agreements governing such cooperation are also known as "Section 810" agreements, after 10 Code of Federal Regulations Part 810.

A nuclear cooperation agreement is not required for transmission of nuclear-related information, except for restricted data. The term "restricted data," as well as other terms used in the statute, is defined in 42 U.S.C. §2014. "Restricted data" means "all data concerning (1) design, manufacture, or utilization of atomic weapons; (2) the production of special nuclear material; or (3) the use of special nuclear material in the production of energy." Restricted data, however does not "include data declassified or removed from the Restricted Data [sic] category" pursuant to the AEA. A nuclear cooperation agreement is necessary, though not necessarily sufficient, to permit the transfer of restricted data.

[3] "Special nuclear material" means (1) plutonium, uranium enriched in the isotopes 233 or 235, and any other material that is determined to be special nuclear material, but does not include source material, or (2) any material artificially enriched by any of the foregoing, but does not include source material.

[4] Nuclear cooperation agreements with nuclear weapon states recognized by the NPT are provided for in the AEA, and are therefore non-exempt agreements. The NPT defines nuclear weapon states as those that exploded a nuclear weapon or other nuclear explosive device prior to January 1, 1967: China, France, Russia, the United Kingdom, and the United States.

[5] The United States has concluded more than 20 bilateral nuclear cooperation agreements, as well as similar agreements with the European Atomic Energy Community and the IAEA. See *Nuclear Commerce: Governmentwide Strategy Could Help Increase Commercial Benefits from U.S. Nuclear Cooperation Agreements with Other Countries*, Government Accountability Office, GAO-1 1-36, November 2010.

[6] See CRS Report RL33016, *U.S. Nuclear Cooperation with India: Issues for Congress*, by Paul K. Kerr. The agreement entered into force in December 2008.

[7] When calculating periods of "continuous session" under the AEA, every calendar day is counted, including Saturdays and Sundays. Only days on which either chamber has adjourned for more than three days pursuant to the adoption a concurrent resolution authorizing the adjournment do not count toward the total. If Congress adjourns its final session *sine die*, continuity of session is broken, and the count must start anew when it reconvenes.

[8] In light of the Supreme Court's 1983 decision in *INS v. Chadha*, passing a concurrent resolution could invite a legal challenge because it is arguably unconstitutional. Although not provided for in the AEA, Congress could choose to pass a joint resolution of disapproval or a bill stating in substance it did not approve.

[9] Section 129b. (2) states that the prohibitions described in the previous section "shall not apply to exports, reexports, transfers, or retransfers of radiation monitoring technologies, surveillance equipment, seals, cameras, tamper-indication devices, nuclear detectors, monitoring systems, or equipment necessary to safely store, transport, or remove hazardous materials ... except to the extent that such technologies, equipment, seals, cameras, devices, detectors, or systems are available for use in the design or construction of nuclear reactors or nuclear weapons."

In: U.S. Nuclear Cooperation with Other Countries
Editor: Gabriel A. Oliveira ISBN: 978-1-61209-910-1

Chapter 2

Nuclear Commerce: Governmentwide Strategy Could Help Increase Commercial Benefits from U.S. Nuclear Cooperation Agreements with Other Countries

United States Government Accountability Office

Why GAO Did This Study

The United States has 26 agreements in force for peaceful nuclear cooperation. Under the U.S. Atomic Energy Act of 1954, as amended, these agreements are a prerequisite to certain aspects of U.S. nuclear cooperation with other cooperating partners. GAO was asked to (1) quantify the amount and value of U.S. nuclear exports facilitated by these agreements, (2) assess U.S. efforts to support the U.S. nuclear industry's ability to compete for sales, and (3) examine U.S. nuclear industry challenges to exporting. To conduct this work, GAO reviewed and assessed data collection efforts by U.S. agencies from 1994 through 2008, analyzed available data, and interviewed U.S. industry representatives and U.S. and foreign government officials.

What GAO Recommends

GAO recommends that Commerce (1) identify additional nuclear data that may better quantify the export benefits of nuclear cooperation agreements, (2) review its strategy document to identify markets and include benchmarks for evaluating progress, and (3) consider ways the interagency trade promotion committee may obtain a comprehensive range of U.S. industry views. Commerce agreed with our first two recommendations but disagreed with the third, stating that it already has mechanisms in place to obtain industry views. GAO is making this recommendation because Commerce needs to strengthen interagency coordination efforts to promote nuclear trade.

WHAT GAO FOUND

No single federal agency systematically tracks and reports the data necessary to determine the amount and value of U.S. nuclear exports facilitated by U.S. nuclear cooperation agreements. Data from the departments of Commerce, Energy (DOE), State, and the Nuclear Regulatory Commission (NRC) contain gaps and in some cases were not sufficiently detailed for GAO's reporting purposes. Using data from the United Nations Commodity Trade Statistics database, GAO found that the United States' share of global exports of nuclear material, reactors, and components has declined in the last 15 years. For example, the amount of U.S. exports of sensitive nuclear material such as natural and enriched uranium remained stable, while the U.S. share of global exports for these materials decreased significantly, from 29 percent to 10 percent, from 1994 through 2008. The United States also imports sensitive nuclear material, nuclear reactors, major components and equipment, and minor reactor parts from other countries. GAO found that in sum, the United States was a net importer of nuclear components and materials, which may indicate a lack of comparative advantage in this industry.

Commerce has an initiative to coordinate interagency efforts and identify and respond to the U.S. nuclear industry's trade policy challenges, but the initiative has made limited progress and does not include a well-defined strategy to support and promote U.S. nuclear industry efforts to compete globally. DOE, NRC, and State officials told us they rely on Commerce to develop and lead U.S. nuclear industry export promotion activities. In October 2008, Commerce established the Civil Nuclear Trade Initiative to help promote the competitiveness of the U.S. nuclear industry. The initiative aims to, among other things, coordinate interagency efforts and identify and respond to trade policy challenges faced by the U.S. nuclear industry. However, the initiative has made limited progress. For example, the initiative's interagency trade promotion committee to coordinate interagency efforts on behalf of U.S. industry has received briefings from only three U.S. nuclear companies; though Commerce officials report many more companies would like to participate. In addition, the initiative's strategy document has some limitations, in that it does not establish goals, does not have an implementation plan, and does not contain metrics for measuring its progress, which are critical for agencies to achieve intended goals.

Commerce, State, and DOE officials as well as U.S. industry representatives identified challenges facing the U.S. nuclear industry, including a decline in domestic manufacturing capabilities, increased international competition, and U.S. industry's liability concerns. In addition, U.S. industry representatives and U.S. foreign government officials GAO interviewed also identified challenges that, in their view, impede the U.S. nuclear industry's ability to compete globally for nuclear trade, including a DOE process for authorizing the transfer of U.S. nuclear technology and technical information overseas. In particular, industry representatives told us they believe that DOE's regulations are outdated and place U.S. companies at a competitive disadvantage.

ABBREVIATIONS

ASME	American Society of Mechanical Engineers
CSC	Convention on Supplementary Compensation for Nuclear Damage of 12 September 1997
CINTAC	Civil Nuclear Trade Advisory Committee
DOE	Department of Energy
EURATOM	European Atomic Energy Community
IAEA	International Atomic Energy Agency
ITA	International Trade Administration
NEI	Nuclear Energy Institute
NMMSS	Nuclear Materials Management and Safeguards System
NPT	Nuclear Nonproliferation Treaty
NRC	Nuclear Regulatory Commission
Paris Convention	Paris Convention on Third Party Liability in the Field of Nuclear Energy
SILEX	Separation of Isotopes by Laser Excitation
TPCC	Trade Promotion Coordinating Committee
UAE	United Arab Emirates
U.N. Comtrade	United Nations Commodity Trade Statistics Database
USTR	United States Trade Representative
Vienna Convention	Vienna Convention on Civil Liability for Nuclear Damage

November 4, 2010

The Honorable Howard L. Berman
Chairman
The Honorable Ileana Ros-Lehtinen
Ranking Member
Committee on Foreign Affairs
House of Representatives

The United States has 26 agreements in force for peaceful nuclear cooperation with foreign countries, the European Atomic Energy Community (EURATOM), the International Atomic Energy Agency (IAEA), and Taiwan.[1] A nuclear cooperation agreement is a bilateral agreement that establishes a framework for civilian nuclear cooperation, including the transfer of certain nuclear material and components of nuclear reactors between cooperating countries. Figure 1 shows the partners with which the United States has, or previously had, a nuclear cooperation agreement in force.

Increasing energy demand and concerns regarding climate change have heightened worldwide interest in nuclear power. IAEA has reported that more than 60 countries are considering nuclear power to help meet their energy needs and estimates that between 10 and 25 new countries will bring their first nuclear power plants online by 2030 as part of an ongoing "nuclear renaissance." By 2030, IAEA estimates the world's capacity for nuclear electricity production will have significantly increased, with most of this increase occurring in countries that have established civilian nuclear power programs, such as China, Japan, and South Korea. China, for example, has announced that it intends to spend $50 billion to build 32 new nuclear plants by 2020. Both India and Pakistan are moving forward with plans to significantly increase their production of nuclear power, building plants that will more than double their production of nuclear energy in the next decade. In addition, countries such as Jordan and Vietnam, which do not yet have civilian nuclear power programs, are actively moving to build the necessary regulatory infrastructure. Other countries, such as Egypt, Indonesia, Libya, and Thailand have expressed their intent to build civilian nuclear power plants. Still others, such as Algeria, Belarus, Nigeria, and Yemen are considering moving forward with civilian nuclear power programs.

These markets potentially represent substantial economic opportunities for the United States and the international nuclear industry. For example, in December 2009 the United Arab Emirates (UAE) selected a consortium led by a South Korean company to build four nuclear power plants, a deal reportedly valued at $20 billion.[2] According to the U. S.-UAE Business Council, a business organization committed to advancing trade between the United States and the UAE, had a U.S. firm been selected, this deal could have potentially generated 10,000 jobs in the United States. The Department of Commerce estimates that every $1 billion in exports by U.S. companies represents 5,000 to 10,000 jobs. President Obama has announced an administration goal of doubling U.S. exports over the next 5 years, and in March 2010 established the National Export Initiative to enhance and coordinate federal efforts to facilitate the creation of U.S. jobs through the promotion of exports.

This report responds to your request that we conduct a review of the export benefits of these nuclear cooperation agreements.[3] Specifically, our objectives were to (1) quantify the amount and value of U.S. nuclear exports facilitated by these agreements from 1994 through 2008, (2) assess U.S. government efforts to support the U.S. nuclear industry's ability to compete for sales facilitated by nuclear cooperation agreements between the United States and other partners, and (3) examine U.S. nuclear industry challenges to exporting, as identified by industry representatives and U.S. and foreign government officials.

To quantify the amount and value of U.S. nuclear exports, we reviewed and assessed data collection efforts by the departments of Commerce, Energy (DOE), and State; the Nuclear Regulatory Commission (NRC); and other U.S. agencies, including the International Trade Commission, Customs and Border Protection, and the Office of the United States Trade Representative (USTR). We obtained and reviewed data from DOE's and NRC's Nuclear Materials Management and Safeguards System (NMMSS), Commerce's Bureau of the Census, and reviewed NRC records on specific licenses for material exports. We determined the value of U.S. exports of nuclear material, reactors, major components and equipment, and minor reactor parts, and determined the United States' relative share of global exports for these commodities, by analyzing data from the United Nations Commodity Trade Statistics Database (U.N. Comtrade) to estimate the value of U.S. exports using other countries' reported U.S. import data.[4] For nuclear reactors, major components and equipment, and minor parts, we analyzed data only under that specific category in the Harmonized System. To assess the reliability of the data from U.N. Comtrade, we reviewed United Nations' records of data evaluation and related documentation and determined that the data were sufficiently reliable for our purposes to estimate exports of nuclear material, reactors, major components and equipment, and minor reactor parts from 1994 through 2008, the most recent year for which complete data were available. To assess U.S. government efforts to support the U.S.

nuclear industry's ability to compete for sales facilitated by nuclear cooperation agreements between the United States and other countries, we analyzed pertinent documentation and interviewed Commerce, State, DOE, and NRC officials to identify key U.S. nuclear export initiatives and activities, and discussed coordination, evaluation, and outcomes of these efforts. Specifically, we reviewed and evaluated Commerce's Civil Nuclear Trade Initiative. We also reviewed NRC export license data to identify suppliers of U.S. nuclear material, reactors, major components and equipment, and minor reactor parts in the last 15 years. To examine challenges to the U.S. nuclear industry, we conducted interviews with a nonprobability sample of seven U.S. or foreign nuclear industry companies and industry representatives with significant business interests in the United States, including companies that obtained NRC-specific licenses to authorize the export of nuclear reactors or major components from 1994 through 2008. We interviewed industry representatives and government officials regarding the challenges they face. Because these industry representative interviews offer opinions regarding the U.S. government's efforts to promote U.S. nuclear technology, they cannot be generalized to the entire universe of civilian nuclear exporters. We reviewed U.S. government regulations including DOE's 10 C.F.R. Part 810 and NRC's 10 C.F.R. Part 110 that govern the process for obtaining authorizations for U.S. persons to engage in the production of special nuclear material in foreign countries, or licenses to export nuclear material and equipment, respectively. We also conducted interviews with officials from several major nuclear importing and exporting countries. Additional details on our scope and methodology can be found in appendix I.

We conducted our work from August 2009 through November 2010 in accordance with generally accepted government auditing standards. Those standards require that we plan and perform the audit to obtain sufficient, appropriate evidence to provide a reasonable basis for our findings and conclusions based on our audit objectives. We believe that the evidence obtained provides a reasonable basis for our findings and conclusions based on our audit objectives.

BACKGROUND

Under Section 123 of the U.S. Atomic Energy Act of 1954, as amended, nuclear cooperation agreements are a prerequisite to certain aspects of civilian U.S. nuclear cooperation with countries and other cooperating partners. The Atomic Energy Act also requires that these agreements include, among other things, guarantees from the partners that they will maintain safeguards over nuclear materials and equipment transferred and adequate physical security for all nuclear material transferred. Nuclear cooperation can include exports of sensitive nuclear material, including special nuclear material such as Plutonium-239, Uranium-235, and Uranium-233, and source nuclear material such as natural uranium; nuclear reactors and their major components; and other activities related to the nuclear fuel cycle.[5] While these agreements provide the framework and authorization for civilian nuclear cooperation, they do not guarantee that cooperation will take place or that material or components will be transferred. Even after a nuclear cooperation agreement has entered into force, a U.S. company seeking to export special nuclear material, reactors, or reactor components must obtain a license from NRC to do so.

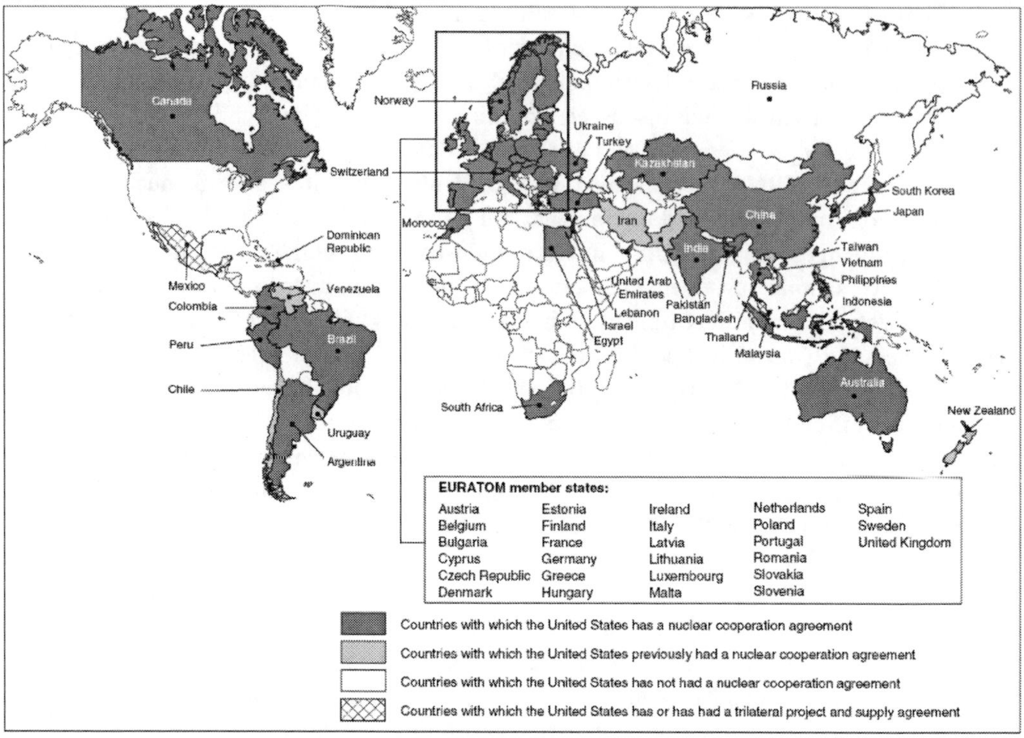

Sources: GAO analysis of Department of State data and Map Resources.

Notes: The United States also has a set of trilateral project and supply agreements with Mexico and IAEA. We included these agreements because they were entered into pursuant to the United States' nuclear cooperation agreement with IAEA. The United States also previously had a trilateral project and supply agreement with Malaysia and IAEA. The United States has an additional agreement with Australia for cooperation for the Separation of Isotopes by Laser Excitation (SILEX) technology for uranium enrichment. For a list of partners with which the United States has a nuclear cooperation agreement in force, see appendix II. In addition, the United States previously had nuclear cooperation agreements with Chile, Dominican Republic, Iran, Israel, Lebanon, New Zealand, Pakistan, Philippines, Uruguay, Venezuela, and Vietnam.

Figure 1. Cooperating Partners with Which the United States Currently Has or Previously Had a Nuclear Cooperation Agreement.

Several U.S. federal agencies develop, implement, and monitor nuclear cooperation agreements and nuclear trade:

- State is responsible for negotiating any proposed nuclear cooperation agreement, with the technical assistance and concurrence of DOE and in consultation with NRC officials. The nuclear cooperation agreement is accompanied by a summary of relevant classified information prepared for the President in consultation with the Director of National Intelligence. State takes the lead in working with country officials to develop the specific terms and conditions that are included.
- DOE has a statutory role in negotiating the agreements, and under Section 57b of the Atomic Energy Act is responsible for authorizing activities that may directly or indirectly assist in the production of special nuclear material outside of the United

States. According to DOE documents, such activities include design information for technology and consulting services related to the production of special nuclear material.[6] DOE has promulgated regulations implementing its authorization process at 10 C.F.R. Part 810, known as "Part 810 authorizations." In addition, DOE, in coordination with NRC, maintains the NMMSS, a database that contains current and historic data on the possession, use, and shipment of nuclear materials, including data on U.S.-supplied nuclear material transactions with other countries and international organizations. DOE, NRC, and other federal agencies primarily rely on NMMSS to track nuclear material exports to foreign countries.

- NRC is responsible for issuing licenses for nuclear exports. The commodities under NRC export licensing authority include: nuclear reactors; major components, equipment, and minor reactor parts; and sensitive nuclear materials such as enriched uranium, plutonium, and depleted uranium. A specific license, which is a license that must be individually approved by NRC, is required for most transfers of special nuclear material, major reactor components, and some minor reactor components.[7] A general license, which is a license that is provided for by rule and which allows exports of certain material without an application to NRC, is required for small quantities of nuclear material and for minor reactor components that do not require a specific license.[8] A NRC license does not guarantee that an export will take place.
- Commerce works to maximize U.S. commercial competitiveness by increasing market access for U.S. businesses and promoting export growth and has a critical role in implementing the Administration's National Export Initiative. Within Commerce, the International Trade Administration (ITA) mission includes: (1) strengthening the competitiveness of U.S. industry, (2) promoting trade and investment, and (3) working to ensure fair trade and compliance with trade agreements. In particular, ITA's Office of Energy and Environmental Industries' primary mission is to facilitate global trade and to support the economic competitiveness of U.S. energy and environmental technologies firms. It provides economic and policy analysis to help U.S. energy companies compete in the global market and works with industry to ensure that its input is reflected in trade and domestic policy development, negotiations, and implementation. In addition, Commerce's Bureau of the Census compiles export statistics on nuclear material, nuclear reactors, major components, and equipment exported. The Bureau of the Census reports this information to U.N. Comtrade database. Finally, Commerce's Bureau of Industry and Security licenses the exports of dual-use items—items that can be used for both civilian and military applications, some of which are nuclear-related items.

The United States has recently signed a number of nuclear cooperation agreements, some of which have raised concerns. For example, the 2008 U.S.-India nuclear cooperation agreement raised concerns that the agreement would negatively impact the international nuclear nonproliferation regime under the Nuclear Nonproliferation Treaty (NPT).[9] The 2009 U.S.-UAE agreement raised concerns based on UAE's past lack of cooperation with international sanctions to prevent Iran from developing nuclear weapons and ballistic missiles. Notwithstanding those concerns, the UAE deal has also been cited as a model for future U.S. nuclear cooperation agreements, as it secured additional nonproliferation commitments from the UAE. In particular, the UAE agreement to forgo domestic enrichment of uranium or reprocessing of spent nuclear fuel without agreement of the parties has alleviated some proliferation concerns.[10] In May 2010, the Administration resubmitted to Congress a peaceful nuclear cooperation agreement with Russia to, among other things, establish a legal framework for DOE to work with Russia on large- scale development of nuclear energy.[11] However, as we reported in June 2009 and September 2010, [12] when the agreement was first submitted to Congress by the Bush Administration on May 13, 2008, Russia's status as a nuclear weapons state, the size of its nuclear complex, and past proliferation concerns—including weaknesses in the Russian export control system—raised concerns.[13]

ALTHOUGH NO FEDERAL AGENCY SYSTEMATICALLY TRACKS DATA TO ASSESS EXPORT BENEFITS OF AGREEMENTS, AVAILABLE INFORMATION INDICATES U.S. SHARE OF GLOBAL NUCLEAR EXPORTS HAS DECREASED SIGNIFICANTLY OVER THE LAST 15 YEARS

No single federal agency systematically tracks and reports the data necessary to determine the amount and value of U.S. nuclear exports facilitated by U.S. nuclear cooperation agreements. However, based on available information, we found that the United States' share of global exports of nuclear material, reactors, and components has declined in the last 15 years.

U.S. Agencies Do Not Systematically Track and Report Data Needed to Assess Export Benefits Facilitated by Nuclear Cooperation Agreements

No single federal agency systematically tracks and reports the data necessary to determine the amount and value of U.S. nuclear exports facilitated by U.S. nuclear cooperation agreements. Existing data from Commerce, DOE, and NRC contain gaps and in some cases were not sufficiently detailed for our reporting purposes. Specifically:

- *Number of reactors, major components, equipment, and minor reactor parts.* No federal agency tracks the number of nuclear reactors and major components exported.[14] While a NRC export license authorizes the potential transfer of a nuclear reactor, major components, or equipment, NRC does not require an exporter to report the actual transfer or export of such commodities. In addition, according to NRC officials, a NRC-issued specific license may be valid for a decade or longer, and a company could export a number of major components under a single license, which complicates attempts to identify whether and when a company exported a particular component. Further, NRC does not have an electronic record- keeping system to track specific licenses for export of nuclear reactors, major components or equipment, or special nuclear material. NRC officials told us that they were working on a system to track licenses but that the only authoritative records of specific licenses are the paper files at NRC. NRC officials told us they informally track licenses with a spreadsheet they estimated to be reliable to approximately the mid-1990s but could not guarantee that it was authoritative. Commerce collects data that report the volume of nuclear reactors, major components and equipment, and minor reactor parts, exported by weight; however, these data do not capture the number of nuclear reactors, major components and equipment, or minor reactor parts exported.

- *Type of major component, equipment, or minor reactor part.* Commerce data do not provide sufficient detail to determine the type of major component, equipment, or minor reactor part exported. While Commerce data capture the value of nuclear reactors and nuclear reactor parts exported, the data on nuclear parts do not differentiate between the components and equipment exported under a NRC specific license and the components and equipment exported under a NRC general license.
- *Value of nuclear material exports.* DOE-NRC's NMMSS data and Commerce data contain gaps regarding the trade value of nuclear material exports. In particular, while NMMSS captures the value of DOE's U.S. government-owned sensitive nuclear material exports, such as highly enriched uranium—which it maintains as classified information—NMMSS does not record data regarding the value of commercially exported material such as low-enriched uranium. NRC and DOE officials told us they do not track the estimated value of exports because this is the proprietary information of exporting companies. Furthermore, DOE has not configured NMMSS to include the World Customs Organization international standard for classification of traded goods, the Harmonized Commodity Description and Coding System. This system, used by Commerce to classify traded goods, including nuclear material, reactors, and components, could assist in determining which NMMSS records on the specific shipments of U.S. special nuclear material exports and NRCspecific licenses correspond to Commerce's value data. In addition, Commerce's data do not record the specific enrichment level of enriched uranium. In particular, Commerce's data do not specify the varying degree of enrichment of low-enriched uranium, which could range from 1 percent of the isotope U-235 to 20 percent of the isotope U-235 in a particular batch of enriched uranium and could have widely varying values. Therefore, we were unable to determine the value of exports of U.S. enriched uranium to other countries from 1994 through 2008 using DOE, NRC, and Commerce data.
- *Volume or value of U.S. nuclear services exported.* We found there are no available data regarding exports of services, which according to Commerce officials is an increasingly important and growing market segment for the U.S. nuclear industry.[15]

The U.S. Share of Global Exports of Sensitive Nuclear Material and Reactors, Major Components, and Minor Parts Has Declined in the Last 15 Years

Based on available data, while the value of U.S. exports of sensitive nuclear material has remained stable from 1994 through 2008, the U.S. share of global exports of sensitive nuclear material declined significantly over the same period. Furthermore, while the value of U.S.

exports of nuclear reactors, major components, and minor parts increased, the U.S. share of global exports declined slightly over this same period.

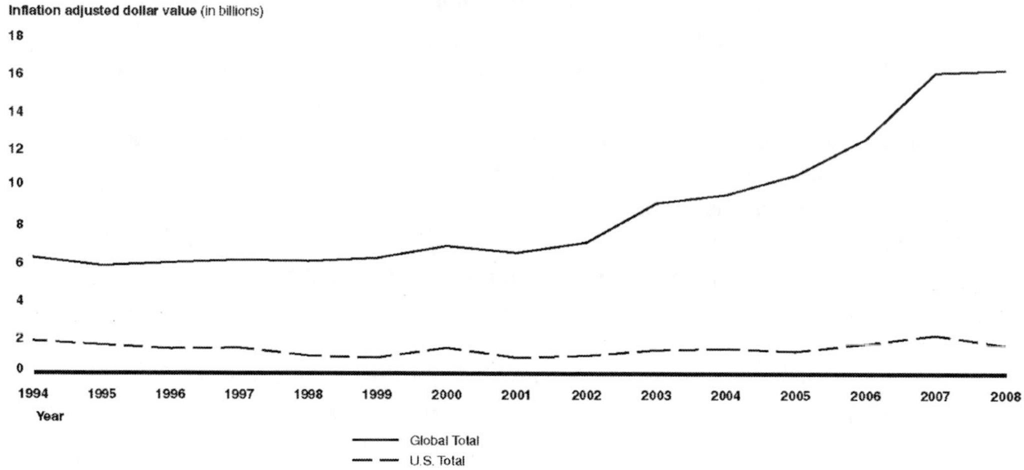

Source: GAO analysis of U.N. Comtrade data.

Figure 2. Comparison of Value of U.S. and Global Exports of Natural Uranium, Enriched Uranium, and Plutonium, 1994 through 2008, in 2010 U.S. Dollars.

Table 1. Top Foreign Customers for U.S. Exported Natural Uranium, Enriched Uranium, and Plutonium from 1994 through 2008, in 2010 U.S. Dollars.

Billions of dollars		
Country	**Total**	**Percentage**
Japan	$12.9	63%
Netherlands	2.1	10
France	1.2	6
South Korea	1.2	6
Germany	1.0	5
Subtotal	18.5	90
Other[a]	2.1	10
Total	**$20.7**	**100%**

Source: GAO analysis of U.N. Comtrade data.

[a] Note: According to U.N. Comtrade, the other countries that the United States exported nuclear material including natural uranium, enriched uranium, and plutonium to include Argentina, Australia, Austria, Bangladesh, Belgium, Brazil, Bulgaria, Canada, China, Colombia, Cyprus, Czech Republic, Denmark, Egypt, Estonia, Finland, Greece, Hungary, Indonesia, Ireland, Italy, Kazakhstan, Latvia, Lithuania, Luxembourg, Malta, Mexico, Morocco, Norway, Peru, Poland, Portugal, Republic of South Africa, Romania, Slovakia, Slovenia, Spain, Sweden, Switzerland, Thailand, Turkey, Ukraine, and the United Kingdom.

U.S. Share of Global Exports of Sensitive Nuclear Material Declined Significantly Over the Last 15 Years

Based on U.N. Comtrade data,[16] we found the value of U.S. exports of sensitive nuclear material such as natural uranium, enriched uranium, and plutonium remained stable, while the U.S. share of global exports for these materials decreased significantly from 1994 through 2008. Specifically, according to U.N. Comtrade data, U.S. exports of natural uranium, enriched uranium, and plutonium were approximately $1.8 billion in 1994 and $1.6 billion in 2008, in 2010 U.S. dollars. However, over this same period, annual global exports of such material more than doubled from $6.2 billion in 1994 to $16.1 billion in 2008, as indicated in figure 2. Effectively, over this 15-year period, the U.S. share of the annual global export market decreased significantly from approximately 29 percent to 10 percent, or from one-third to one-tenth of the market. Global exports of sensitive nuclear material remained relatively unchanged until approximately 2002, when global exports increased rapidly from $7 billion in 2002, to $9.5 billion in 2004, and to $12.5 billion in 2006.

U.N. Comtrade data also indicate that from 1994 through 2008, the United States exported approximately $20.7 billion, in 2010 U.S. dollars, in natural uranium, enriched uranium, and plutonium to countries with which the United States has a nuclear cooperation agreement. As indicated in table 1, Japan was the United States' largest foreign customer for sensitive nuclear material, and accounted for an estimated $12.9 billion, approximately 63 percent, of U.S. sensitive nuclear material exports.

U.S. Share of Global Exports of Nuclear Reactors, Major Components and Equipment, and Minor Reactor Parts Decreased Over Last 15 Years

Based on U.N. Comtrade data, we found that although the value of U.S. exports of nuclear reactors, major components and equipment, and minor reactor parts increased during the period from 1994 through 2008, the U.S. share of global exports of such goods has decreased. Specifically, according to U.N. Comtrade data, the value of U.S. exports of nuclear reactors, major components and equipment, and minor reactor parts increased from approximately $160 million in 1994 to $308 million in 2008, while annual global exports of these commodities nearly tripled from $1.5 billion in 1994 to $4.3 billion in 2008, in 2010 U.S. dollars, as indicated in figure 3. Because U.S. exports did not increase at the same rate as global trade, the United States' share of the global export market for these goods effectively declined from approximately 11 percent to 7 percent. Over time, global exports of nuclear goods have increased steadily, experiencing a slight dip in 2006 to $2.6 billion, but recovering by 2008 to $4.3 billion.

We found the United States exported approximately $4.4 billion, in 2010 U.S. dollars, of nuclear reactors, major components and equipment, and minor reactor parts to countries with which it had a nuclear cooperation agreement. According to our analysis, the United States' top customers for nuclear reactors, major components and equipment, and minor reactor parts were Japan, South Korea, Mexico, Spain, and the Czech Republic. As shown in table 2, these five countries accounted for $3.1 billion, or approximately 70 percent in 2010 U.S. dollars, of all of U.S. exports of nuclear reactors, major components and equipment, and minor parts.

Table 2. Top Foreign Customers for U.S. Exported Nuclear Reactors, Major Components and Equipment, and Minor Reactor Parts from 1994 through 2008, in 2010 U.S. Dollars.

Millions of dollars		
Country	**Total**	**Percentage**
Japan	$920	21%
South Korea	683	16
Mexico	532	12
Spain	492	11
Czech Republic	434	10
Subtotal	3,062	70
Other[a]	1,324	30
Total	**$4,386**	**100%**

Source: GAO analysis of U.N. Comtrade data.

[a] Note: According to U.N. Comtrade, the other countries that the United States exported nuclear reactors, major components and equipment, and minor reactor parts to include Argentina, Australia, Austria, Bangladesh, Belgium, Brazil, Canada, China, Colombia, Cyprus, Denmark, Egypt, Finland, France, Germany, Indonesia, Ireland, Italy, Kazakhstan, Morocco, Netherlands, Norway, Peru, Poland, Republic of South Africa, Romania, Slovakia, Slovenia, Sweden, Switzerland, Thailand, Turkey, Ukraine, and United Kingdom.

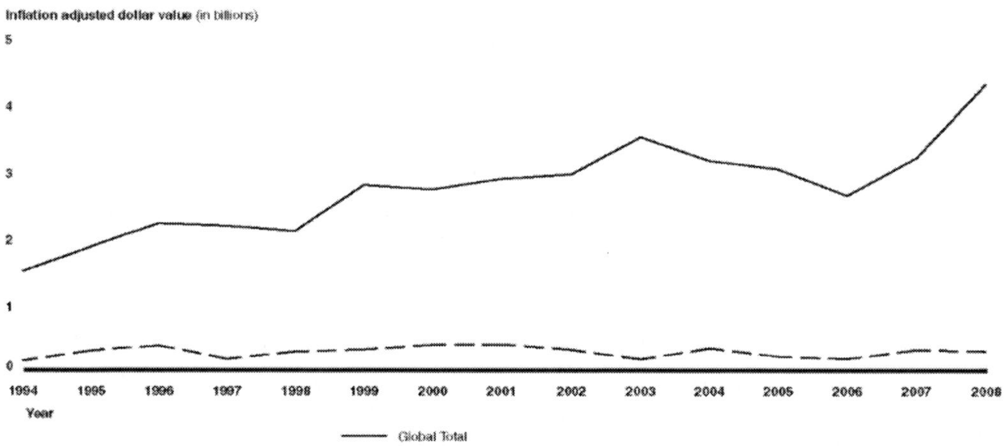

Source: GAO analysis of U.N. Comtrade data.

Figure 3. Comparison of Value of U.S. and Global Exports of Nuclear Reactors, Major Components and Equipment, and Minor Reactor Parts, 1994 through 2008, in 2010 U.S. Dollars.

In addition, our review of NRC records of specific licenses indicates that U.S. companies did not participate in the majority of new reactor construction projects overseas during this period, consistent with the declining share of the United States in global nuclear trade. Specifically, NRC issued eight licenses authorizing the export of nuclear reactors and major components from 1994 through 2008: three licenses for projects in South Korea, three for projects in China, one for Colombia, and one for Thailand. In addition, six licenses issued before 1994 were extended for use as of July 2010 for projects in Brazil, Slovenia, South Korea, and Taiwan. For purposes of comparison with the global market, according to World Nuclear Association data, 61 civilian nuclear power reactors in 14 foreign countries began operating from 1994 through 2008.[17]

Finally, the United States imports sensitive nuclear material, nuclear reactors, major components and equipment, and minor reactor parts from other countries. U.N. Comtrade data indicate that the United States imported $39.2 billion in natural uranium, enriched uranium, and plutonium, in 2010 U.S. dollars, from 1994 through 2008. The United States imported the greatest amount of these materials from Russia, at a cost of approximately $11.3 billion. During this period, the United States also imported approximately $1.2 billion in nuclear reactors, major components and equipment, and minor reactor parts from other countries. The United States imported the greatest amount of these nuclear reactors, components, and equipment from Sweden, at a cost of approximately $362.7 million. In sum, the United States was a net importer of nuclear components and materials, which may indicate a lack of comparative advantage in this industry.

Commerce Has an Initiative to Coordinate Interagency Efforts, but Has Made Limited Progress and Does Not Have a Well- Defined Strategy to Promote the U.S. Nuclear Industry Globally

DOE, NRC, and State officials told us they rely on Commerce to develop and lead U.S. nuclear industry export promotion activities because Commerce's mission includes providing trade support to U.S. businesses and trade promotion. Commerce launched the Civil Nuclear Trade Initiative in 2008 to help promote the competitiveness of the U.S. nuclear industry; however, this initiative has made limited progress. According to Commerce officials, the initiative aims to identify the U.S. civil nuclear industry's most pressing trade policy challenges and most promising commercial opportunities and coordinate public and private sector efforts to address them in a way that supports the industry's endeavors to rebuild its manufacturing base.

Specifically, the initiative, as outlined in Commerce's export promotion strategy document, consists of the following four major activities:

- *Interagency nuclear trade coordination committee.* In 2008, under the Trade Promotion Coordinating Committee (TPCC), Commerce formed the TPCC Civil Nuclear Trade Working Group, an interagency nuclear industry subcommittee, to coordinate interagency efforts related to commercial nuclear trade. Membership is drawn from Commerce, State, DOE, among other agencies. Commerce officials told us that the TPCC Civil Nuclear Trade Working Group meets about four times a year, when there are pertinent issues to discuss. Commerce, DOE, NRC, and State officials told us that the meetings are a useful forum to promote interagency dialogue and inform agency officials of ongoing efforts to assist the U.S. nuclear industry. As of June 2010, Commerce officials told us the TPCC Civil Nuclear Trade Working Group had received briefings from three companies that construct civilian nuclear power reactors and reactor components. However, Commerce officials said five other smaller companies who supply nuclear materials and services had expressed an interest in briefing the TPCC Civil Nuclear Trade Working Group, but had not yet had an opportunity to do so. In July 2010, Commerce decided to hold monthly meetings of the TPCC Civil Nuclear Trade Working Group to increase the interaction between U.S. industry and the interagency community.

- *Nuclear trade industry advisory committee.* Commerce established the Civil Nuclear Trade Advisory Committee (CINTAC) to obtain industry advice on enhancing U.S. nuclear industry competitiveness. Members are drawn from the U.S. nuclear industry. Established in October 2008, Commerce officials told us CINTAC did not meet for more than 16 months because of delays associated with the change in U.S. administrations, and associated delays in administrative guidance on selecting and confirming members. According to Commerce officials, as of August 2010, CINTAC had held three formal meetings and formed five subcommittees. Some of the issues raised by the committee include (1) strengthening the U.S. commercial nuclear manufacturing industry through tax credits, loan guarantees, and other incentives; (2) streamlining U.S. export controls; and (3) pursuing the Convention on Supplementary Compensation for Nuclear Damage.[18] CINTAC's charter expired in September 2010, requiring a new charter and appointment of new members. A June 2, 2010, letter from the Chairman of CINTAC to the Secretary of Commerce stated that CINTAC's subcommittees intended to make recommendations before the end of its charter, but requested that the charter be renewed to give the group sufficient time to add to the recommendations it is developing on behalf of the U.S. nuclear industry. CINTAC made the following two recommendations to the Secretary of Commerce: (1) that Commerce request the Office of the United States Trade Representative (USTR) conduct an accelerated assessment of what the committee called uncompetitive trade practices in the global commercial nuclear industry and (2) that Commerce work with USTR to issue an interim report within 6 months focused on the global civilian nuclear industry. A USTR official told us that the agency had reviewed CINTAC's letter to the Secretary of Commerce and was discussing with Commerce officials and CINTAC committee members how best to address the needs of the industry group. In addition, on September 16, 2010, CINTAC made its final recommendations to the Secretary of Commerce. CINTAC recommended, among other things, (1) creating a team of U.S. industry representatives, government officials, and other experts to focus on supporting nuclear opportunities domestically and internationally, and leveling the international playing field; and (2) establishing a policy in coordination with DOE, State, and USTR to encourage international companies that are owned, controlled, or subsidized by their national governments to eliminate unfair competitive structures or financing.
- *Stakeholder resources.* The initiative's stakeholder resources consist of a May 2009 online Civil Nuclear Exporters Guide that presents information generally publicly available on the Internet, including overviews of U.S. export licensing procedures and contact information for key U.S. agency export control offices. Commerce officials told us they were in the process of updating the guide. Commerce has also participated in export control seminars and presented briefings to U.S. companies. Future plans include: (1) developing criteria for prioritizing best prospect markets for

U.S. nuclear industry, (2) developing in-depth export control workshops for U.S. industry, (3) improving the quality of data from current collection efforts, and (4) providing a more thorough analysis of existing data on the economic benefits accrued from U.S. civilian nuclear exports.

- *Trade promotion.* Commerce officials said trade promotion activities have consisted of organizing trade missions, participating in trade policy discussions, organizing industry outreach events, and developing a joint trade promotion declaration to promote the development of civilian nuclear programs worldwide. Commerce officials told us they had organized one trade mission on behalf of the U.S. nuclear industry to Eastern Europe—a July 2010 trip to the Czech Republic, Poland, and Slovakia—and two U.S. nuclear trade promotion missions to IAEA. Commerce officials said they have also certified, supported, and participated in U.S. Chamber of Commerce- and non-U.S. government- organized trade missions. Commerce officials told us they are planning future trade missions to target additional countries, and that they will host a U.S. nuclear industry promotion event at the 2010 IAEA General Conference.[19] Commerce also participated in nuclear trade policy discussions with delegations visiting the United States from Japan, Lithuania, Mongolia, Poland, and Thailand. In addition, Commerce officials have organized two industry outreach events to increase engagement between U.S. government officials and industry representatives. According to Commerce, approximately 150 to 200 industry representatives attended these outreach events. Furthermore, Commerce and DOE have jointly developed and signed joint trade promotion declarations with Italy and Poland that encourage nuclear industry entities to seek opportunities to participate in the construction of nuclear power plants and the provision of related supporting infrastructure and services. The declarations include commitments by the partner countries that they will develop transparent procedures for awarding contracts within their nuclear energy industries and in a way that does not unduly favor companies from countries that may, according to Commerce officials, directly subsidize their nuclear industry.

We found that the Commerce initiative's strategy document has some limitations. We have reported in the past that it is important to align strategic goals with strategies for achieving those goals.[20] Specifically, we have noted that defining the mission and desired outcomes, measuring performance, and using performance information to identify performance gaps, are critical if agencies are to be accountable for achieving intended results.[21] However, the initiative's strategy document does not contain any of these steps or measures. Specifically, the strategy document has not identified or established key market opportunities and goals, does not have an implementation plan, and does not include metrics, benchmarks, or timelines to measure progress in meeting specific goals.[22] For example, although the strategy document includes future plans for each of the four major activities, these plans are vaguely outlined and do not set target dates for their completion. Furthermore, the strategy document does not identify or prioritize countries as targets for U.S. promotion efforts. Commerce officials acknowledged that, while they believe the initiative is aligned with Commerce and ITA goals, they were working to more clearly state timelines, desired outcomes, milestones, and success metrics for each initiative activity. Commerce officials also told us that one staff member is engaged in a market prioritization study, which has a planned issuance date of late 2010.

Commerce officials told us the initiative is being undertaken with limited resources and without dedicated funding. Specifically, at present, the initiative is being organized and run by one staff member working at 80 percent time and 2 staff members working at 75 percent and 50 percent, respectively. Commerce officials reported that the initiative is supported mainly through the budget of ITA's Office of Energy and Environmental Industries and other offices when funds are available.

INDUSTRY REPRESENTATIVES AND U.S. AND FOREIGN GOVERNMENT OFFICIALS REPORTED THAT THE U.S. NUCLEAR INDUSTRY FACES MANY CHALLENGES COMPETING GLOBALLY

Officials from Commerce, State, and DOE, and foreign governments, as well as U.S. nuclear industry representatives told us there are multiple challenges for U.S. companies competing globally for nuclear trade. First, officials and industry representatives told us that the U.S. nuclear industry may not be well-positioned to secure the trade benefits facilitated by nuclear cooperation agreements and U.S. companies face increased international competition from foreign state-owned suppliers, which are heavily subsidized and supported by their governments. Second, officials and industry representatives told us that in their view U.S. industry is at a competitive disadvantage due to the lack of a global liability regime. In addition, U.S. agency and industry representatives and foreign government officials we interviewed identified challenges that U.S. companies face that, in their view, impede the U.S. nuclear industry's ability to compete globally.

U.S. Nuclear Industry May Not Be Well-Positioned to Secure Trade Benefits and U.S. Companies Face Competition from State- Owned Nuclear Firms

Officials from Commerce, State, and DOE, as well as U.S. industry told us they are concerned that the U.S. nuclear industry may not be well- positioned to secure the trade benefits facilitated by nuclear cooperation agreements due to a decline in domestic manufacturing capabilities, increased international competition, and U.S. industry's liability concerns. In January 2010, Commerce's ITA reported that the U.S. nuclear industry has atrophied, and according to U.S. government officials and nuclear industry representatives, may lack the capability to manufacture certain components and equipment needed to produce large civilian power reactors.[23] Further, a State Department official testified in May 2010 that it would be difficult for the United States to build its own nuclear reactor without importing a number of significant components. As we noted on page 17 of this report, the United States is a net importer of nuclear components and materials. While Commerce officials acknowledged that U.S. nuclear manufacturing capability has eroded, they said that the domestic supply chain is showing signs of revitalization. In particular, Commerce officials noted that while the number of facilities certified by the American Society of Mechanical Engineers (ASME) to produce commercial nuclear-grade components had fallen from 440 in the 1980s to 120 by the early 2000s, the number of facilities had increased to 255 as of mid-2008, which they took as evidence of potential recovery.

Moreover, government and industry officials told us that the nature of the civilian nuclear power market has changed dramatically. According to the Nuclear Energy Institute (NEI), the main nuclear industry advocacy organization, U.S. reactor designs and U.S. reactor manufacturers and fuel suppliers had the dominant market share for several decades.[24] However, beginning in the 1970s, and steadily increasing since then, vendors of nuclear reactors in Europe and Asia developed their own reactor designs and capacity to manufacture components and fuel. By the 1980s, buyers of power reactors could choose among reactor suppliers from many countries, including Canada, France, Japan, and Russia. Russian officials told us its Ministry of Foreign Affairs is aggressively seeking to sign as many nuclear cooperation agreements as possible with an eye to expanding into new markets.

Commerce noted that U.S. firms face formidable competition abroad from foreign companies that receive strong financial and political support through direct government ownership or subsidies. In addition, foreign governments may place greater emphasis on supporting bids through high- level advocacy or by providing customers additional services and expertise. For example, according to media reports, the President of France and President of South Korea recently traveled to UAE to advocate for their country's respective bids to build new reactors in the UAE. In addition, French officials told us that their government's philosophy on nuclear cooperation includes providing a package of regulatory, financial, and technical assistance to partner countries developing their civilian nuclear power program.

Commerce officials told us that some of the largest markets for nuclear goods and services, such as France, Russia, and South Korea, have significant barriers to entry for U.S. companies because of the presence of a state-owned competitor. According to Commerce, of the 61 civilian nuclear reactors outside the United States that began operating from 1994 through 2008, 18 reactors—almost 30 percent—went into operation in France, Russia, and South Korea—countries with their own state-owned nuclear companies.

U.S. Companies Challenged By Lack of Global Liability Regime

U.S. government officials and industry representatives have reported that the lack of a comprehensive global liability regime hampers U.S. industry's ability to secure civilian nuclear contracts and places U.S. firms at a competitive disadvantage. Currently, there are two conventions applicable to liability for nuclear accidents; the United States is not a party to either one.[25] Participation in the conventions by other countries is also relatively limited. A third convention, CSC, has been developed to supplement the other two and, according to IAEA's International Expert Group on Nuclear Liability, it serves as an umbrella agreement to create a single global liability regime. The United States has ratified this convention, but it has not yet come into force. U.S. industry representatives told us that without a global liability regime in force that channels liability for accidents at a nuclear facility to the operator of that facility, they fear that they may be held liable as suppliers. By contrast, a U.S. industry representative told us that foreign companies that are state-owned may not face the same problem because they may be indemnified by their government. Furthermore, in the absence of a global liability regime, U.S. industry representatives told us that they cannot obtain insurance sufficient to cover their potential liabilities resulting from a potential nuclear reactor accident overseas. Industry representatives told us that if they export to a nation without strong domestic or international liability laws that channel responsibility to the operator of the civilian nuclear facility, their company could be sued into bankruptcy as a supplier. Representatives from several U.S. companies told us they do not export to countries without liability protection.

Commerce officials said that the U.S. nuclear industry faces significant challenges in competing to win contracts for new civilian nuclear power reactors outside the United States because of fierce competition from foreign companies that may receive stronger financial and political support from their governments, which creates an uneven playing field. One major U.S. nuclear industry company told us the French and South Korean nuclear industries benefit from high-level support in their foreign marketing activities, which can only give them an advantage compared to U.S. companies. Many of the U.S. company representatives we spoke with, as well as representatives from the NEI, told us that unless the U.S. government develops a governmentwide strategy and greater commitment to its nuclear industry, the United States' ability to secure export benefits from nuclear cooperation agreements in the form of reactor and large component and equipment sales is uncertain. In addition, a representative from a major U.S. nuclear industry company told us that the company's future innovation depends on its ability to participate in the current global marketplace. The U.S. nuclear industry will not be able to sustain research and development programs of any size without participating fully in the global nuclear renaissance, according to this company.

Some U.S. Policies and Practices Are Viewed by Industry Officials and Foreign Governments as Impediments to U.S. Nuclear Exports

U.S. industry representatives and foreign government officials we interviewed identified certain U.S. government policies and practices that, in their view, impede the U.S. nuclear industry's ability to compete globally for nuclear trade. In particular, industry representatives told us that DOE's regulations governing the Part 810 authorization process place U.S. companies at a competitive disadvantage.[26] In May 2010, a representative from NEI stated that the current Part 810 rules have the unintended effect of standing in the way of cooperative programs and information exchange.[27] U.S. nuclear industry representatives told us that DOE's regulations governing DOE's process to authorize U.S. company representatives to engage in the production of special nuclear material outside the United States and implement its Atomic Energy Act responsibilities lack clarity and are outdated. In particular, these representatives told us that DOE's Part 810 authorization regulations are vaguely defined and that DOE interprets its authority to include transfers of technology and technical assistance too broadly.[28] DOE has not published any guidelines to clarify the 810 authorization process for civilian nuclear exporters; however, DOE officials said they regularly hold meetings and give presentations at agency and industry events to explain Part 810 regulations. A DOE document states that the agency makes a broad and comprehensive interpretation of the regulation to include the provision of technology in the form of assistance or services to any nuclear power program outside the United States. [29] In addition, Part 810 regulations require a specific authorization to engage directly or indirectly in the production of special nuclear material in 77 countries listed in 10 C.F.R. § 810.8; however, DOE officials acknowledged that the list was out of date. For example, countries with which the United States currently has nuclear cooperation agreements—such as Kazakhstan, Ukraine, and the UAE—are still listed.

Officials from several U.S. companies told us that nuclear firms in other countries do not face this additional step to authorize the transferring of technology and technical information. DOE officials said that the Part 810 authorization regulations are outdated and cumbersome and that they were working on revising the regulations to make them more clear and efficient, but could not estimate when revised regulations would be completed.

Finally, in one instance, U.S. government practices had the unintended consequence of negatively affecting another country's decision to work with U.S. companies. A senior foreign country official told us that the U.S. government's inability to work cooperatively had influenced that country's decision to purchase civilian nuclear power reactor fuel from a non-U.S. supplier. The senior official expressed frustration that that the United States did not treat the country more as a business partner. A June 2010 unclassified State communication summarizing our interview noted that the senior foreign country official stated that because buying fuel from the United States was not very "client friendly," the partner country's nuclear power plants chose U.S. competitors to supply their fuel.

CONCLUSIONS

NRC, DOE, and Commerce are making efforts to collect some information regarding nuclear exports. However, agencies' efforts are piecemeal, and no federal agency collects or tracks information on exports to fully assess the amount and value of exports facilitated by U.S. nuclear cooperation agreements. We believe that the lack of an integrated governmentwide approach that links the potential growth of such agreements and the growing worldwide nuclear market to a well-defined strategy to support and promote U.S. nuclear exports is negatively affecting U.S. nuclear industry's ability to compete globally. Given that hundreds of new civilian nuclear power reactors are expected to be constructed in foreign countries in the coming decades, the lack of a reliable U.S. government assessment of the U.S. nuclear industry's exports and competitive position may impede the ability of the industry to make a meaningful contribution to the President's goal of doubling national exports in 5 years. Moreover, the absence of a comprehensive strategy to support the U.S. nuclear industry may result in the relevant agencies, in particular Commerce, not being equipped to provide targeted and informed support to U.S. industry to help it compete for sales of civilian nuclear power reactors overseas. While a comprehensive strategy is not a panacea to improve the competitiveness of the declining U.S. nuclear industry—ultimately U.S. industry must provide quality products and services at a competitive price—such a strategy, as well as improved industry data collection, could be an important step in helping promote our nuclear industry.

Further, Commerce's 2-year effort to develop an export promotion strategy document has produced limited results. While the outline of such a strategy appears to be envisioned by Commerce in its Civil Nuclear Trade Initiative, the initiative has yet to identify key market opportunities for U.S. industry, establish goals, and develop an implementation plan with timelines and metrics to measure progress toward these goals. The absence of such information in its strategy means that Commerce and other agencies may not be devoting their resources to the markets and activities that show the greatest promise for U.S. companies. Furthermore, while the formation of the TPCC Civil Nuclear Trade Working Group as part of the initiative is a positive step, we are concerned that the group has heard from only three U.S. nuclear exporting firms and not from smaller firms that supply nuclear materials and services.

There may be opportunities to provide more immediate support to assist the U.S. nuclear industry's efforts to become more competitive globally. Representatives from U.S. nuclear companies stated that they face an outdated and unclear DOE Part 810 authorization process that impedes their ability to quickly provide potential customers with the types of technology, information, and technical assistance needed to compete for potential sales. Industry officials also said that they lack guidance in interpreting the DOE Part 810 process, which causes confusion in the Part 810 application process. Furthermore, the lack of an updated 810.8 list of countries may create an unnecessary hurdle for a U.S. company in obtaining permission to provide potential customers with the types of technology and design information that they may require in selecting a vendor for a new civilian reactor.

RECOMMENDATIONS FOR EXECUTIVE ACTION

To help federal agencies gain a better understanding of how—and the extent to which—nuclear cooperation agreements impact exports to other countries, we recommend that the Secretary of Commerce, working with the Secretaries of Energy and State and the Chairman of the Nuclear Regulatory Commission, take the following three actions:

- identify what additional nuclear export data and information may be necessary to better quantify the export benefits associated with these agreements;
- review, with an eye toward strengthening, Commerce's existing nuclear export promotion strategy document to, among other things, identify key market opportunities for U.S. nuclear industry, and develop key goals and an implementation plan for achieving these goals; and
- consider ways for the TPCC Civil Nuclear Trade Working Group to obtain a more comprehensive range of U.S. industry views.

We are also making one recommendation to the Secretary of Energy. Specifically, we recommend that, while considering the broader revisions to the Part 810 regulations as planned, the Secretary of Energy

- review the current Part 810 authorization process; develop guidelines to help clarify the types of technology, information, and technical assistance that require a Part 810 authorization; and consider whether some countries should be removed from the Part 810.8 list to facilitate U.S. exports to countries with which the United States has a nuclear cooperation agreement in force.

APPENDIX I. OBJECTIVES, SCOPE, AND METHODOLOGY

We addressed the following questions during our review: (1) quantify the amount and value of U.S. nuclear exports facilitated by these agreements; (2) assess U.S. government efforts to support the U.S. nuclear industry's ability to compete for sales made possible by nuclear cooperation agreements between the United States and other countries; and (3) examine U.S. nuclear industry challenges to exporting as identified by industry representatives and U.S. and foreign government officials.

To quantify the amount and value of U.S. nuclear exports, we reviewed and assessed data collection efforts by the Department of Commerce (Commerce), the Department of Energy (DOE), the Nuclear Regulatory Commission (NRC), the Department of State (State), and other U.S. agencies including the International Trade Commission, Customs and Border Protection, and the Office of the United States Trade Representative. To determine the volume of U.S. nuclear material and number and amount of nuclear reactor component and equipment exports facilitated by nuclear cooperation agreements, we obtained and reviewed data on U.S. exports of low-enriched uranium, highly enriched uranium, and plutonium, from DOE and NRC's Nuclear Materials Management and Safeguards System (NMMSS), Commerce's Bureau of the Census, and reviewed NRC records on specific licenses for material exports. To determine the value of U.S. exports of nuclear material, nuclear reactors, major components and equipment, and minor reactor parts, and to determine the United States' relative share of global exports of these commodities, we obtained and analyzed United Nations Commodity Trade Statistics Database (U.N. Comtrade) data to estimate the value of U.S. exports using other countries' reported U.S. import data. For nuclear reactors, major components and equipment, and minor parts, we analyzed data only under that specific category in the Harmonized System. We considered the trade of all sensitive nuclear material and minor reactor parts transferred to a country with which the United States has a nuclear cooperation agreement to be facilitated by agreement, regardless of whether a nuclear cooperation agreement would be needed. To assess the reliability of the data from U.N. Comtrade, we reviewed the United Nations records of data evaluation and related documentation and determined that the data were sufficiently reliable to estimate exports of nuclear material, nuclear reactors, major components and equipment, and minor reactor parts from 1994 through 2008, the most recent year for which complete data were available. Because the United States has a trilateral project and supply agreement with Mexico and the International Atomic Energy Agency (IAEA), which was entered into pursuant to the United States' nuclear cooperation agreement with IAEA, we included Mexico in our analysis. It was outside the scope of this review to account for the value of technical and engineering services in the United States that facilitate the export of nuclear materials, including uranium mining, enrichment, and fuel fabrication services. It was also outside the scope of this review to account for the full scope of facilities, components, and equipment related to the full nuclear fuel cycle. To determine with which countries the United States has nuclear cooperation agreements, we obtained and reviewed the texts of U.S. nuclear cooperation agreements from the United Nations Treaty Collection database. We interviewed DOE, NRC, State, and Commerce officials regarding the results of our data analysis.

To assess U.S. government efforts to support the U.S. nuclear industry's ability to compete for sales made possible by nuclear cooperation agreements between the United

States and other countries, we interviewed Commerce, State, DOE, and NRC officials to identify key U.S. nuclear export initiatives and activities, and discussed coordination, evaluation, and outcomes of these efforts. Specifically, we reviewed and evaluated Commerce's Civil Nuclear Trade Initiative. We reviewed U.S. government regulations including DOE's 10 C.F.R. Part 810 and NRC's 10 C.F.R. Part 110 that govern the process for obtaining authorizations for U.S. persons to engage in the production of special nuclear material in foreign countries, or licenses to export nuclear material and equipment respectively. We reviewed NRC export license data to identify suppliers of U.S. nuclear material, nuclear reactors, and key components and equipment in the last 16 years and found there were three such companies—two other license holders were merged into another license holder.

To examine challenges to the U.S. nuclear industry identified by industry representatives and U.S. and foreign government officials, we conducted interviews with a nonprobability sample of seven U.S. and foreign nuclear industry companies and industry representatives with significant business interests in the United States, including companies that obtained NRC specific licenses to authorize the export of nuclear reactors or major components from 1994 through 2008. Specifically, we interviewed representatives from AREVA, Curtiss Wright Corporation, GE Hitachi Nuclear Energy, The Babcock & Wilcox Company, USEC, Inc., and Westinghouse, as well as with officials from trade organizations including the Nuclear Energy Institute. We interviewed foreign government officials regarding U.S. government efforts to promote U.S. nuclear technology, equipment, fuel, and fuel services. Because these industry representative interviews offer opinions regarding the U.S. government's efforts to promote U.S. civilian nuclear industry exports and major challenges they face, they cannot be generalized to the entire universe of civilian nuclear exporters. We also interviewed DOE, NRC, State, and Commerce officials regarding challenges they view with regard to the U.S. nuclear industry's ability to compete for sales made possible by nuclear cooperation agreements going forward.

We conducted our work from August 2009 through November 2010 in accordance with generally accepted government auditing standards. Those standards require that we plan and perform the audit to obtain sufficient, appropriate evidence to provide a reasonable basis for our findings and conclusions based on our audit objectives. We believe that the evidence obtained provides a reasonable basis for our findings and conclusions based on our audit objectives.

APPENDIX II. LIST OF PARTNERS WITH WHICH THE UNITED STATES HAS A NUCLEAR COOPERATION AGREEMENT

As of January 1, 2010, bilateral Agreements for Cooperation were in force with the following partners:

- Argentina
- Australia[30]
- Bangladesh
- Brazil

- Canada
- China
- Colombia
- European Atomic Energy Community (EURATOM)
- Egypt
- India
- Indonesia
- International Atomic Energy Agency (IAEA)
- Japan
- Kazakhstan
- Morocco
- Norway
- Peru
- South Korea
- South Africa
- Switzerland
- Taiwan[31]
- Thailand
- Turkey
- Ukraine
- United Arab Emirates

In addition, the United States has a trilateral project and supply agreement with Mexico and the IAEA.

APPENDIX III. COMMENTS FROM THE DEPARTMENT OF COMMERCE

Note: GAO comments supplementing those in the report text appear at the end of this appendix.

UNITED STATES DEPARTMENT OF COMMERCE
The Under Secretary for International Trade
Washington, D.C. 20230

OCT 21 2010

Mr. Glen Aloise
Director, Natural Resources and Environment
U.S. Government Accountability Office
441 G Street, N.W.
Washington, DC 20548

Dear Mr. Aloise:

Thank you for the opportunity to comment on the draft Government Accountability Office (GAO) report titled "*Nuclear Commerce: Government-Wide Strategy Could Help Increase Commercial Benefits from U.S. Nuclear Cooperation Agreements with Other Countries.*" In general, we agree with most of your recommendations. We are pleased that the GAO acknowledged the role Commerce's Civil Nuclear Trade Initiative can play in realizing U.S. commercial benefits from the global expansion of nuclear power. I have enclosed comments relevant to Commerce's efforts referenced in the draft report.

The Obama Administration intends to continue its efforts to support the safe and secure expansion of nuclear power for peaceful purposes and promote U.S. nuclear industry leadership worldwide. We have achieved some success and are committed to making further progress.

Sincerely,

Francisco J. Sánchez

Enclosure

INTERNATIONAL TRADE ADMINISTRATION

See comment 1.
See comment 2.
See comment 3.
See comment 4.

Department of Commerce Technical Comments on
Nuclear Commerce: Government-Wide Strategy Could Help Increase Commercial Benefits from U.S. Nuclear Cooperation Agreements with Other Countries
GAO-11-36 Draft Report
October 14, 2010

- Highlights Page, second paragraph: Please include the complete objective of the U.S. Department of Commerce's Civil Nuclear Trade Initiative (CNTI): "Identify the U.S. civil nuclear industry's most pressing trade policy challenges and most promising commercial opportunities and coordinate public and private sector efforts to address them in a way that supports the industry's endeavors to rebuild its manufacturing base."

As indicated in staff discussions with GAO and in our previous written submissions, including the above, we believe that the CNTI has made substantial progress since its launch two years ago. In our view, reference to "limited progress" does not accurately depict activities to date.

We believe CNTI has a well-defined strategy to meet its intended objectives, which includes the following:

(1) Refocus the U.S. Government (USG) efforts on civil nuclear trade as well as nuclear nonproliferation, rather than exclusively on nuclear nonproliferation as was the focus for the past thirty years.
(2) Develop a leadership role for the International Trade Administration (ITA) within the USG and with U.S. industry on commercial nuclear issues. Before the CNTI was launched in 2008, ITA did not have any presence within the USG or with the U.S. nuclear industry on promoting nuclear trade.
(3) Establish formal and informal mechanisms to ensure the U.S. civil nuclear industry and USG are informing and advising one another on commercial and policy issues surrounding the global expansion of nuclear power.
(4) Using information from USG agencies and the U.S. nuclear industry, identify the most significant trade barriers that are holding U.S. companies back from benefiting from the global resurgence of nuclear power and develop and implement strategies to eliminate those barriers. Conversely, identify the best prospect markets for U.S. civil nuclear technology and services and prioritize ITA's nuclear trade policy and promotion activities to reflect those rankings.
(5) Provide resources to stakeholders that will advance their understanding of, and ability to compete in, the evolving international civil nuclear market.

While it is correct that there have been only three companies so far that have formally briefed the Trade Promotion Coordinating Committee's interagency working group on civil nuclear trade, there have been numerous occasions where we have organized company-interagency briefings. For example, the 17 U.S. nuclear companies that made up the Civil Nuclear Trade Advisory Committee (CINTAC) under its inaugural charter briefed the interagency civil nuclear trade working group multiple times (in both open

See comment 5.
See comment 6.
See comment 7.
See comment 8.
See comment 7.

and closed sessions). Informally, we organize regular briefings and meetings for industry and USG officials to discuss relevant nuclear trade issues. Briefings have covered such topics as Jordanian, Indian, British, and Brazilian nuclear market overviews; the Nuclear Regulatory Commission's International Regulatory Development Program; the International Atomic Energy Agency's (IAEA) Convention on Supplementary Compensation for Nuclear Damage (CSC); and small modular reactors.

Additionally, since 2008, ITA has organized, with interagency support, two major industry outreach events, each attended by approximately 150-200 industry representatives. Both of these events aimed to increase engagement between government and industry on topics of mutual interest such as trade promotion, export financing, export controls, manufacturing, labor, and supply chain constraints, nuclear cooperation agreements, and indemnification in the case of nuclear accident.

We recognize that a comprehensive document with goals, an implementation plan, and metrics for measuring CNTI progress is needed, and we are in the process of writing such a document. It is incorrect, however, to assume that in the absence of such a formal, public document, we have not set goals, crafted an implementation plan, or established how we measure success. We have, in fact, incorporated these important components of program management in our CNTI work. We have also worked to ensure that the CNTI's program of work is consistent with, and supports the achievement of, ITA's long-term strategic goals, including: (1) advancing U.S. international and commercial strategic interests; (2) enhancing U.S. competitiveness in domestic and international markets; (3) broadening and deepening the U.S. exporter base; and (4) identifying and resolving unfair trade practices.

Using the five goals of the CNTI strategy identified in the first paragraph of this response, we would like to highlight aspects of our strategy and implementation plan for achieving those goals and how we measure success.

Goal (1) Refocusing the USG interagency on civil nuclear trade as well as nuclear nonproliferation rather than exclusively on nuclear nonproliferation: The strategy and implementation plan is to ensure that the commercial consequences for the U.S. nuclear industry is a topic on the agenda at key interagency nuclear meetings and that ITA staff participate in such discussions to make the case for considering commercial issues. Further, our strategy is to encourage government agencies that did not support nuclear projects, (e.g., U.S. Trade Development Agency and Overseas Private Investment Corporation), to do so, and those that do support nuclear projects (e.g., EXIM Bank), to do more. Also worth noting here is that before the CNTI, nuclear trade issues were not given the proper attention that they are now given at U.S. Trade Representative (USTR). Success is measured by ITA being consulted on all issues that could affect the competitiveness of the industry and by agencies becoming more responsive to industry's needs. For example, USTDA has reversed its no-nuclear policy and has actively been organizing civil nuclear reverse trade missions with ITA's support. In another success, USTR has become more engaged on nuclear trade issues, such as actively seeking input

See comment 8.
See comment 7.
See comment 7.
See comment 9.

from civil nuclear companies for information to include in the 2011 National Trade Estimate Report on Foreign Trade Barriers.

Goal (2) Developing a leadership role for ITA within the USG and with U.S. industry on commercial nuclear issues where ITA previously had no role: The strategy and implementation plan includes having the Secretary of Commerce establish a Trade Promotion Coordinating Committee (TPCC) Civil Nuclear Trade Working Group, and inviting USG counterparts to identify a staff member to participate in this working group. The primary purpose of this group is not necessarily to receive individual company briefings. While such briefings are useful, the real utility of the group is to coordinate USG agency efforts in support of the U.S. nuclear industry. ITA chairs the working group and articulates the U.S. nuclear industry's priorities, based on ITA's regular interaction with the industry. The group communicates regularly over email and phone in addition to formal meetings. Specifically, the group seeks to identify which of the industry's greatest trade challenges and opportunities that the USG should, and can, best address in a coordinated fashion. The group has identified bringing into force the IAEA's CSC as its top priority, as industry has identified the lack of a global nuclear liability regime as one of its greatest barriers to entering new markets. The USG is currently encouraging key countries to ratify CSC. The strategy included a January 2010 U.S. interagency-industry mission to key countries to promote CSC as well as follow-up letters to key officials into those countries to encourage CSC ratification. Next steps include an early-2011 CSC promotion seminar in Asia. The strategy also includes inserting CSC language in bilateral commercial declarations signed with key countries as demonstrated by the joint declarations on commercial nuclear cooperation signed with Italy and Poland recently. While the group members communicate on a regular basis, it will begin holding formal monthly meetings given the increasing number of issues that need to be discussed. U.S. nuclear companies will be invited to brief the group for the first half of the two-hour meeting.

The TPCC working group also identified leading and supporting civil nuclear trade missions as one of the best ways to address best prospect markets for industry. This resulted in a July 2010 interagency-supported civil nuclear trade mission to Poland, Czech Republic, and Slovakia, led by Under Secretary of Commerce Francisco Sánchez. These three countries were prioritized under the CNTI through formal criteria-based analyses, as best prospect markets given upcoming tenders and commercial developments. Success is measured by the active participation of the interagency community in working group discussions and activities to identify export opportunities and challenges and develop and implement commercial strategies, and also leverage existing commercial programs, to advance the global competitiveness of the U.S. civil nuclear industry, i.e., the CSC strategy and trade missions.

Goal (3) Establishing formal and informal mechanisms to ensure the U.S. civil nuclear industry and USG are informing and advising one another on the commercial and policy issues surrounding the global expansion on nuclear power: The strategy and implementation plan includes establishing a Civil Nuclear Trade Advisory Committee (CINTAC), which is made up of industry executives representing U.S. nuclear companies

See comment 7.
See comment 9.
See comment 8.
See comment 7.
See comment 8.
See comment 9.

across the supply chain. CINTAC advises the Secretary of Commerce and USG agencies, through the Secretary's role as chair of the TPCC Civil Nuclear Trade Working Group. In particular, CINTAC provides advice and recommendations on USG policies and programs affecting the competitiveness of the U.S. nuclear industry. CINTAC submitted its final recommendations in September 2010, and those recommendations will drive the work of CNTI. An informal mechanism was also established that includes organizing regular briefings by the USG for industry on best prospect markets, as well as USG initiatives and activities that are being developed and implemented to improve industry's ability to compete abroad. CNTI has built an industry database that currently includes over 350 industry representatives which is used for outreach purposes. Success is measured by the quality of industry recommendations and input into trade policy development, negotiations, and implementation.

Goal (4) Using information from the USG and U.S. nuclear industry to identify the most significant trade barriers that are holding U.S. companies back from benefiting from the global resurgence of nuclear power, and developing and implementing strategies to eliminate those barriers. Conversely, identifying the best prospect markets for U.S. nuclear technology and services and prioritizing ITA's nuclear trade policy and promotion work to reflect those priorities: The strategy and implementation plan includes polling industry on top trade barriers, discussing these identified issues within the interagency community, and methodically deciding the trade policy and promotion activities to undertake to ensure maximum effectiveness. Those actions included first-of-their-kind Joint Declarations on Commercial Nuclear Cooperation with both Italy and Poland that demonstrate high-level USG support for commercial nuclear cooperation. These Declarations are an attempt at leveling the playing field for U.S. companies as they compete against state-owned enterprises abroad that regularly enjoy high-level government support. Future Declarations are being explored in other key markets. The importance of other countries adopting U.S. nuclear codes and standards was also identified as a priority issue. In response, ITA organized a U.S.-Brazil Nuclear Codes and Standards Workshop in Brazil in August 2010, and is planning another one in Vietnam for 2011. Both countries were identified as best prospect markets for the U.S. nuclear industry by CNTI's original methodology. Official USG advocacy is also an important part of our trade promotion and policy work. Advocacy is coordinated out of ITA's Advocacy Center in coordination with the goals of the CNTI and with the support of the interagency community. Trade missions and the U.S. industry promotion program organized at the IAEA General Conference are also important components of our trade policy and promotion work. An additional civil nuclear trade mission is planned for September 2011. The measure of success is ultimately the number of contracts signed as a result of these programs and policies, recognizing the long lead-time for nuclear projects and the infancy of the CNTI. Other contributing factors to consider are the impact on the USG's ability to shape policies and regulations in foreign markets; for example, ensuring that codes and standards are consistent with our own so that U.S. businesses have a stronger foothold from which to be able to conduct business in these markets.

See comment 7.
See comment 7.
See comment 9.
See comment 10.
See comment 1.
See comment 11.
See comment 12.

See comment 13.
See comment 14.
See comment 14.

Goal (5) Providing resources to stakeholders that will advance their understanding of, and ability to compete in, the evolving international nuclear market: The strategy and implementation plan included developing an on-line guide to assist U.S. nuclear companies navigate the often complex export controls process. This guide is being currently updated. The Department of Commerce, in partnership with the Nuclear Energy Institute and other USG agencies, organized an Export Controls Workshop in 2009 for U.S. industry new to exporting civil nuclear technology and services. A nuclear trade web portal is currently under development, which will serve as the virtual interface of CNTI and as a clearinghouse of information related to U.S. and international nuclear commerce targeted to both a U.S. and foreign audience. An original report on the competitiveness of the U.S. small modular reactor manufacturers will soon be published and plans are underway to continue publishing reports on timely topics. Additionally, efforts are underway to revise the criteria we developed for prioritizing best prospect markets, with the intention to share the criteria with industry and other U.S. Government agencies, and to continue to use the best prospects analysis to direct the work of CNTI. Improving the clarity and availability of nuclear trade data, as well as analyzing the U.S. economic benefits of present and future nuclear exports, are planned for 2011. Success here is measured by the quality, accuracy, and timeliness of information provided, which contribute to enhancing the global competitive position of U.S. industry.

- Pages 10 and 11: Census Bureau should be consulted to confirm that the summary is correct. You should also note that nuclear services are not captured in any available data, but that services are an increasingly important and growing market segment for U.S. nuclear companies.

- Pages 17-21: Please incorporate the information on CNTI as described above in this response document, as appropriate.

- Page 20, paragraph 2: Please correct the statement regarding number of staff dedicated to CNTI to read, "…one staff member working at 80 percent time, and two staff members working at 75 percent and 50 percent respectively."

- Page 22, paragraph 3: Recommend deleting the statement, "In January 2010, Commerce's ITA reported that the U.S. nuclear industry has atrophied," or correcting it to read, "Commerce's ITA reported that the U.S. nuclear industry has atrophied over the last thirty years, yet has been showing positive signs of new growth." Note this point is also made on page 23, paragraph 1.

- Page 23, paragraph 1: Please add the word "potential" in front of "recovery" in last sentence, and add the following sentence: "In addition, Commerce officials noted that new nuclear manufacturing facilities are being built in the United States."

- Page 23, paragraph 3: Add "and political" between "financial support."

- Page 25, paragraph 2: Add "and political" between "financial support."

See comment 15.
See comment 9 and 4.
See comment 16.

- Page 28, paragraph 1: For clarity, we recommend that you de-link your analysis and conclusions on increasing commercial benefits from the growth of U.S. nuclear cooperation agreements, which are very specific and limited, and instead describe your analysis and conclusions in the context of a growing worldwide nuclear market. Also, it should be recognized that while a comprehensive strategy is important for ITA to support the U.S. nuclear industry, there are limits in terms of what ITA can do to support the industry, given ITA has no regulatory or statutory authority over the issues affecting the industry, nor does ITA have dedicated funding for its nuclear undertakings.
- Page 28, paragraph 2: It should be noted that the production of a strategy document has not been the focus of Commerce's CNTI for the past two years. The CNTI has identified key market opportunities and trade policy challenges for U.S. industry, established goals, implemented strategies to achieve those goals, and currently is assessing the progress toward those goals. A comprehensive strategy document is a good step in moving CNTI forward and we are therefore in the process of drafting such a document that will be released in 2011. Regarding the statement on the TPCC Civil Nuclear Trade Working Group, please refer to the information on this topic as described in the beginning of this response document for clarification on the purpose and role of the working group, and why it should not be evaluated in terms of how many individual, formal company briefings the group had.
- Page 29, last bullet: We recommend deleting this statement because as noted at the beginning of this response document, the TPCC Civil Nuclear Trade Working Group has obtained a broad range of U.S. industry views, including from second tier suppliers and small and medium-sized companies. One example is the multiple briefings the working group has received from the Civil Nuclear Trade Advisory Committee (CINTAC), which has 17 senior industry executives representing all sizes of companies from across the supply chain and located throughout the United States. If you prefer to keep the statement, then it should be clarified, as there is no "nuclear industry Trade Promotion Coordination Committee."

The following are GAO's comments in response to the Department of Commerce's letter dated October 21, 2010.

GAO Comments

1. We modified the text on page 18 of the report to state that the initiative "aims to identify the U.S. civil nuclear industry's most pressing trade policy challenges and most promising commercial opportunities and coordinate public and private sector efforts to address them in a way that supports the industry's endeavors to rebuild its manufacturing base," in response to Commerce's comment. However, we left the original, more general phrasing on the Highlights page intact because it accurately reflects the primary objective of the initiative.
2. We disagree with Commerce's comment that its initiative has made substantial progress. Our report assesses the progress that Commerce has made for each of the initiative's four major activities. We continue to believe that, thus far, Commerce's initiative has made limited progress. In particular, as we note in our report, the TPCC Civil Nuclear Trade Working Group has only heard from three companies, and has not heard from another five that have expressed an interest in briefing it; CINTAC did not hold its first meeting until more than 16 months after it was chartered and its charter has since expired; the initiative's stakeholder resources consist of information

generally available on the Internet, participation in export control seminars, and briefings to U.S. companies; and Commerce has organized one trade mission and a limited number of other industry outreach events.

3. We disagree with Commerce's comment that it has a well-defined strategy to meet intended objectives. We found that the Commerce initiative's strategy document has some limitations, in that the strategy document has not identified or established key market opportunities and goals; does not have an implementation plan; and does not include metrics, benchmarks, or timelines to measure progress in meeting specific goals. Commerce's written comments provide new and clarifying information regarding the key goals of the initiative, which we believe is a positive development. However, it has not provided new information regarding identifying key markets and an implementation plan.
4. Commerce presents new information that the CINTAC industry advisory group briefed the TPCC Civil Nuclear Trade Working Group multiple times. We believe interaction between the TPCC Civil Nuclear Trade Working Group and the CINTAC industry advisory group is a positive development. However, as our report noted, CINTAC's charter expired in September 2010, requiring a new charter and appointment of new members. Commerce's comments did not indicate that CINTAC's charter has been renewed, which casts doubt on the ability of CINTAC to keep the TPCC Civil Nuclear Trade Working Group apprised of market developments. In addition, Commerce's comments stated that it has organized regular briefings and meetings for industry and U.S. government trade officials to discuss, among other things, Jordanian, Indian, British, and Brazilian nuclear market overviews; the CSC, and small modular reactors, and that these meetings are informal. Because Commerce describes these briefings as informal, we believe they do not represent the kind of sustained, regular interaction to coordinate interagency efforts related to civil nuclear trade that the TPCC Civil Nuclear Trade Working Group was designed to provide. We are making this recommendation because Commerce needs to strengthen interagency coordination efforts to promote nuclear trade. Moreover, as our report notes, many of these U.S. company representatives we spoke with, as well as representatives from the Nuclear Energy Institute, told us that unless the U.S. government develops a governmentwide strategy and greater commitment to its nuclear industry, the United States' ability to secure export benefits from nuclear cooperation agreements in the form of reactors, large components, and equipment sales is uncertain.
5. We added the following language to page 20 of the report: "In addition, Commerce officials have organized two industry outreach events to increase engagement between U.S. government officials and industry representatives. According to Commerce, approximately 150 to 200 industry representatives attended these outreach events."
6. See comment 3.
7. Commerce presented its five goals, which is a new characterization of its initiative. However, much of the information presented in this section detailing what Commerce calls "aspects" of its implementation plan contains information already included in our draft report. For example, we discuss CSC; the civil nuclear trade mission to Poland, the Czech Republic, and Slovakia; CINTAC; the joint

declarations on commercial nuclear cooperation; and the export controls workshop on pages 24, 20, 18-19, 20, and 19 of our report, respectively. Therefore, we made no changes to the report in response to this comment.

8. In some cases, the information Commerce presents as "aspects" of its implementation under the newly constituted five goals for the Initiative is new information. For example, Commerce notes in August 2010 it organized a U.S.-Brazil Nuclear Codes and Standards Workshop in Brazil, and another is planned for Vietnam in 2011. In other cases the new information provided by Commerce includes assertions that Commerce did not provide documents to support. For example, Commerce comments that the United States Trade and Development Agency has reversed a "no-nuclear" policy and has, according to Commerce, actively begun organizing reverse civil trade nuclear missions. In addition, Commerce comments that prior to the initiative, "nuclear issues were not given the proper attention that they are now given at USTR." We did not modify the text in response to these comments.
9. In its written comments, Commerce presented new information regarding how it intends to measure the performance of its strategy, after it has more fully developed the strategy document as we recommend and as the agency states it intends to do. Commerce's efforts to provide some information regarding the performance measures should be included in a formal strategy document, as we recommend.
10. We requested data on exports from various Commerce officials on multiple occasions, which Commerce officials were unable to provide in full. Although Commerce officials did not make Census officials available at our meetings or final review, we developed our methodology with and had our results reviewed by economists in GAO's Applied Research and Methods group. Our results were additionally reviewed by Commerce, Energy, State, and NRC officials. However, we did modify the text on pages 11-12 to note that we found there are no available data regarding exports of services, which according to Commerce officials is an increasingly important and growing market segment for the U.S. nuclear industry. We note that Commerce did not provide documents to support its assertion regarding the growth in the U.S. nuclear services market.
11. We modified the text on page 21 to reflect Commerce's updated staffing figures, to show one staff member working at 80 percent time and 2 staff members working at 75 percent and 50 percent, respectively, on the initiative.
12. The source of the statement that Commerce seeks clarification for is the U.S. Department of Commerce's 2010 Energy Industry Assessment ("2010 Assessment"), as noted in footnote 23. Commerce's 2010 Assessment does not state that the U.S. nuclear industry is showing positive signs of growth. However, we modified the text on page 22 to state that the report notes U.S. companies have participated in the world market, often as a minority partner, and invested in research and development for the next generation of reactors. We note on page 22 of the report that, according to Commerce officials, the number of facilities certified by the American Society of Mechanical Engineers (ASME) to produce commercial nuclear-grade components has risen from 120 by the early 2000s, to 255 as of mid-2008, which Commerce officials take as evidence of recovery.

13. We have modified the text on page 23 to note that Commerce officials believe the recent increase in the number of facilities certified by the American Society of Mechanical Engineers (ASME) to produce commercial nuclear-grade components is evidence of "potential recovery," not "recovery," as the report had previously stated.
14. We have modified the text on page 23 and 24 to state that Commerce officials said that U.S. firms face formidable competition abroad from foreign companies that may receive stronger financial and political support.
15. We modified the text to add the words "and the growing nuclear worldwide market" on page 27. Our report notes on page 21 the limited staff and funding available for the Commerce initiative, and also notes that the U.S. government relies on Commerce to promote U.S. nuclear industry exports.
16. We disagree with Commerce's recommendation that we delete our third recommendation that it consider ways for the TPCC Civil Nuclear Trade Working Group to obtain a more comprehensive range of U.S. industry views. We believe it is important to keep this recommendation. As our report notes, thus far only three companies have briefed the TPCC Civil Nuclear Trade Working Group. However, in response to Commerce's comments we modified the text to refer to the TPCC interagency nuclear trade coordination committee as the TPCC Civil Nuclear Working Group in the body of the report and in the third recommendation.

APPENDIX IV. COMMENTS FROM THE DEPARTMENT OF ENERGY

Note: GAO comments supplementing those in the report text appear at the end of this appendix.

See comment 1.

See comment 2.

See comment 3.

See comment 4.

Department of Energy
National Nuclear Security Administration
Washington, DC 20585
October 21, 2010

Mr. Gene Aloise
Director
Natural Resources and Environment
Government Accountability Office
Washington, DC 20548

Dear Mr. Aloise:

The National Nuclear Security Administration (NNSA) appreciates the opportunity to review the Government Accountability Office's (GAO) draft report, GAO-11-36, *NUCLEAR COMMERCE: Government-Wide Strategy Could Help Increase Commercial Benefits from U.S. Nuclear Cooperation Agreements with Other Countries.* We understand that GAO was asked to (1) quantify the amount and value of U.S. nuclear exports facilitated by these agreements from 1994 through 2008, (2) assess U.S. efforts to support the U.S. nuclear industry's ability to compete for sales facilitated by nuclear cooperation agreements, and (3) examine U.S. nuclear industry challenges to exporting as identified by industry representatives and U.S. and foreign government officials.

NNSA generally agrees with the report. While the auditors have done a thorough job, NNSA would like to offer the following comments to clarify points in the draft report.

Page 1: The number of 123 Agreements does not include the SILEX Agreement in that total. If you include the SILEX Agreement, then there are 26 Agreements in force.

Page 6: The second sentence in the Background section states, "The Atomic Energy Act also requires these agreements to include guarantees from the partners that they will maintain safeguards over nuclear materials and equipment transferred and adequate physical security for all nuclear material transferred." This sentence leaves out all the other nonproliferation conditions required in Section 123, such as peaceful uses and no retransfer without prior consent. We recommend citing all such conditions or including a qualifier in the sentence as drafted to state that their reference is an illustrative list.

Page 7: Second bullet should reaffirm the point noted above that the Department of Energy (DOE) has a statutory role in the negotiation of all 123 Agreements. You may want to reference that DOE's role stems from Section 57(b) of the Atomic Energy Act.

Page 26: The GAO has cited industry officials as criticizing Part 810 as standing in the way of commerce without offering a counterview regarding the need for DOE to regulate such technology transfers. We believe that it should be noted that the 810 question was not formally raised with DOE officials nor were DOE officials made aware of specific problems. If industry were to specify what the existing problems are, these could be reviewed to determine whether the issues can be corrected or process improved.

NNSA agrees with the recommendation that in revising the Part 810 regulations we should consider whether some countries should be removed from the Part 810.8 list to facilitate U.S. exports to countries with which the United States has a nuclear cooperation agreement in force.

If you have any questions related to this response, please contact JoAnne Parker, Director, Office of Internal Controls, at 202-586-1913.

Sincerely,

Gerald L. Talbot, Jr.
Associate Administrator
for Management and Administration

cc: Deputy Administrator for Defense Nuclear Nonproliferation

The following are GAO's comments in response to the Department of Energy's letter dated October 21, 2010.

GAO Comments

1. We modified the text on page 1 to state the number of agreements to 26 and added clarifying language on pages 2 and 36 of the draft report.
2. We added the words "among other things" on page 5 of the text to clarify.
3. We modified the text on page 6 to state that DOE has a statutory role in negotiating nuclear cooperation agreements, and we moved existing language that DOE derives its authority from Section 57b of the Atomic Energy Act from a footnote on page 7 to the report body in response to this comment.
4. We disagree with DOE's comment that we did not formally raise the Part 810 question with DOE officials. Because we recognize the importance of the Part 810 regulations, which DOE promulgated pursuant to the authority granted to it by section 57b of the Atomic Energy Act and which aims to ensure that activities associated with the development or production of any special nuclear material outside the United States will not be inimical to U.S. interests, we felt that it was important for DOE to know about industry uncertainty with regard to the requirements of the regulations. On August 11, 2010, we met with DOE and shared preliminary findings regarding Part 810, including specific concerns that the U.S. nuclear industry representatives had shared with us. We discussed these concerns with DOE officials, and those officials acknowledged that the Part 810.8 list was out of date and that DOE has not published any guidelines to clarify the 810 authorization process for civilian nuclear exporters. In addition, DOE officials told us that industry representatives had expressed concerns with the Part 810 authorization

process to DOE for several years, and that DOE intended to revise its regulations to make them more clear and efficient.

End Notes

[1] EURATOM is composed of the 27 countries of the European Union. IAEA, an independent international organization based in Vienna, Austria, is affiliated with the United Nations and has the dual mission of promoting the peaceful uses of nuclear energy and verifying that nuclear materials intended for peaceful purposes are not diverted to military purposes. IAEA had 151 member states as of December 2009. Governmental relations between the United States and Taiwan were terminated on January 1, 1979. All agreements concluded with the Taiwan authorities prior to January 1, 1979, are administered by the American Institute in Taiwan, a nonprofit District of Columbia corporation.

[2] Westinghouse Electric, a U.S.-based nuclear technology supplier, was selected as a part of the consortium to provide instrumentation, control systems, technical design, and engineering support.

[3] In a separate report to be issued early next year, we plan to assess DOE's efforts to monitor and evaluate the security of nuclear material transferred under nuclear cooperation agreements.

[4] Economists generally agree that import data are generally more accurate than export data, due to countries' interest in tracking imports for tariffs and other fees.

[5] The nuclear fuel cycle uses uranium in different chemical and physical forms. This cycle typically includes the following stages: uranium mining, conversion, enrichment, deconversion, fuel fabrication, use of fuel in reactors, interim storage, recycling, and final disposal. Nuclear technology and assistance may be exported without a nuclear cooperation agreement. For example, U.S. companies can obtain what is known as a DOE "Part 810 authorization" to share technical information regarding the company's product specifications and capabilities when bidding on new reactor tenders in foreign countries without a nuclear cooperation agreement in place. In addition, U.S. companies can obtain a license from the Nuclear Regulatory Commission to export certain small quantities of nuclear material and minor reactor components to foreign countries without a nuclear cooperation agreement in place. Also, dual-use items—items that can be used for both civilian and military applications—can be exported without a nuclear cooperation agreement.

[6] According to DOE officials, such items are those not covered by a NRC license. DOE's responsibility derives from section 57b of the Atomic Energy Act, as amended, which prohibits U.S. persons from directly or indirectly engaging or participating in the development or the production of special nuclear material outside the United States, except as specifically authorized by a nuclear cooperation agreement or upon authorization by the Secretary of Energy after a determination that such activity will not be inimical to the interest of the United States.

[7] A specific license is issued to a named person and is effective upon approval of an application by NRC and issuance of licensing documents to the applicant. Major reactor components requiring a specific license include reactor pressure vessels, reactor control rod systems, and primary coolant pumps. Minor reactor components that may require a specific license include reactor pressure tubes, zirconium tubes, reactor internals, and reactor control rod drive mechanisms. A specific license is required for transfer of minor reactor components unless the export is to a country listed in the agency's regulations as approved to receive minor components under a general license.

[8] A general license is effective without the filing of an application with NRC or the issuance of licensing documents to a particular person. Items approved for export under a general license may not be exported to countries that are embargoed, or, in certain cases, restricted. Minor reactor parts may only be exported under a general license to countries listed in NRC regulations. Exports of minor reactor parts to other countries not listed must be under a specific license.

[9] India has not signed or ratified the NPT, under which, among other things, parties undertake to pursue negotiations in good faith on effective measures relating to nuclear disarmament, and nonnuclear weapon state parties undertake not to manufacture or otherwise acquire nuclear weapons or seek or receive any assistance in the manufacture of nuclear weapons or other nuclear explosive devices, and to accept IAEA safeguards on all source or special fissionable material in all peaceful nuclear activities in their territories or under their jurisdiction or control so that IAEA can verify that its nuclear programs are not being used for nuclear weapons purposes.

[10] Uranium enrichment and reprocessing spent nuclear fuel are the nuclear activities of greatest proliferation concern. Uranium enrichment can produce low-enriched uranium for nuclear reactor fuel or highly enriched uranium, which can be used as fissile material in nuclear weapons. Reprocessing spent nuclear fuel separates plutonium from that fuel. Plutonium can also be used as fissile material in nuclear weapons.

[11] Pursuant to Section 123 of the Atomic Energy Act, the President submits the text of a proposed agreement to the Committee on Foreign Relations of the Senate and the Committee on Foreign Affairs of the House of

Representatives for consultation for a period of 30 days of continuous session. The proposed agreement is then to be submitted to Congress (and referred to the above mentioned Committees) for a period of 60 days of continuous session, during which the committees consider it and report recommendations. An agreement that contains all the required elements may be brought into force after expiration of 90 days of continuous session, unless a joint resolution disapproving the agreement has been enacted during that period.

[12] See GAO, *U.S.-Russia Nuclear Agreement: Interagency Process Used to Develop the Classified Nuclear Proliferation Assessment Needs to Be Strengthened*, GAO-09-743R (Washington, D.C.: June 30, 2009) and *GAO, 2010 Resubmission of the U.S.-Russia Nuclear Cooperation Agreement: Further Actions Needed by State and Other Agencies to Improve the Review of the Classified Nuclear Proliferation Assessment*, GAO-10-1039R (Washington, D.C.: Sept. 21, 2010).

[13] The Bush Administration withdrew the agreement from congressional review in September 2008 because of the conflict between Russian and Georgia that erupted in the prior month.

[14] Civilian nuclear reactors are generally assembled at the site where they will be operating. Major components for the reactors are shipped separately. Therefore, the export of a "nuclear reactor" does not mean the export was of a fully assembled nuclear reactor, but instead likely indicates export of its major components.

[15] Commerce did not provide documents to support its assertion regarding the growth in the U.S. nuclear services market.

[16] In part because of the data limitations we identified earlier in this report, we used U.N. Comtrade data to estimate U.S. exports by totaling other countries' reported imports from the United States. Economists generally agree that import data are generally more accurate than export data, due to countries' interest in tracking imports for tariffs and other fees. For example, in 2002 the U.S. Commerce Foreign Trade Division reported that, because of the volume of trade with Canada, the United States derives its statistics on exports to Canada from import data reported by Canada.

[17] According to Commerce officials, 8 of these 61 reactors went into operation in India, and construction for these reactors was started before the U.S.-India nuclear cooperation agreement. An additional 11 reactors were constructed in China, some of which began construction before the U.S.-China nuclear cooperation agreement was registered in 2004; and since that time Westinghouse, a U.S.-based nuclear company, has had success there. In addition, 18 reactors came into operation from markets that are de facto closed to U.S. company participation because of the presence of a state-owned competitor: namely, in France, Russia, and South Korea. Finally, an additional 11 came online in Japan, at least 2 of which, according to Commerce officials, were designed by a consortium, which included a U.S. reactor vendor.

[18] The United States has ratified the Convention on Supplementary Compensation for Nuclear Damage of 12 September 1997 (CSC). At the time of our review, the CSC had not entered into force because it had not yet met the threshold of participation with ratification by at least five countries with a combined capacity of at least 400,000 units of installed nuclear capacity.

[19] IAEA's General Conference meets annually, typically in September, to consider and approve the agency's program and budget and to decide on other matters brought before it by the Board of Governors, the Director General, or member states. The IAEA's 53rd General Conference in 2009 included over 1,400 delegates from IAEA member states.

[20] See GAO, *Managing for Results: Critical Issues for Improving Federal Agencies' Strategic Plans*, GAO/GGD-97-180 (Washington, D.C.: Sept. 16, 1997).

[21] See GAO, *Executive Guide: Effectively Implementing the Government Performance and Results Act,* GAO/GGD-96-118, (Washington, D.C.: June 1996).

[22] GAO has reported that developing mechanisms to monitor, evaluate, and report on the results of an effort, and using strategic and annual performance plans to establish goals and strategies to reinforce agency accountability can enhance and sustain collaborative efforts by agencies. See GAO, *Results-Oriented Government: Practices that Can Enhance and Help Sustain Collaboration Among Agencies*, GAO-06-15, (Washington, D.C.: Oct. 21, 2005).

[23] U.S. Department of Commerce, *2010 Energy Industry Assessment* (Washington, D.C., January 2010). However, the Commerce report noted that U.S. companies have participated in the world market, often as a minority partner, and invested in research and development for the next generation of reactors.

[24] NEI is responsible for establishing unified nuclear industry policy on all matters affecting the U.S. nuclear industry.

[25] These conventions are the Vienna Convention on Civil Liability for Nuclear Damage (Vienna Convention) and the Paris Convention on Third Party Liability in the Field of Nuclear Energy (Paris Convention). The Joint Protocol Relating to the Application of the Vienna Convention and the Paris Convention of 21 September 1988 links these two conventions. According to State officials, the United States cannot become party to these other conventions because the U.S. nuclear liability regime does not require channeling of legal liability for nuclear damage to the facility operator, as these conventions require.

[26] DOE is responsible for authorizing activities that may directly or indirectly assist in the production of special nuclear material outside of the United States. According to DOE documents, such activities include design information for technology and consulting services related to the production of special nuclear material.

[27] Hearing on the Future of U.S.-International Nuclear Cooperation, House Committee on Foreign Affairs, Subcommittee on Terrorism, Nonproliferation, and Trade, May 6, 2010.

[28] According to 10 C.F.R. Part 810, DOE requires a "specific" Part 810 authorization to provide sensitive nuclear technology for an activity in any foreign country, or to engage in or provide assistance or training in, among other things, designing, constructing, fabricating, operating, or maintaining facilities for the enrichment or reprocessing of sensitive nuclear material, the production of heavy water, and production reactors. DOE in Section 810.3 defines a production reactor as a nuclear reactor specially designed or used primarily for the production of plutonium or uranium-233. Furthermore, U.S. law places additional requirements on transfers to India pursuant to the Henry J. Hyde United States-India Peaceful Atomic Energy Cooperation Act.

[29] DOE considers technology to cover technical assistance and technical data for the purposes of 810 authorizations. Technical assistance may take forms such as instruction, skills, training, working knowledge, consulting services, or any other assistance as determined by the Secretary of Energy. Technical assistance may involve transfer of technical data. Technical data may take forms such as blueprints; plans; diagrams; models; formulae; engineering designs, specifications, manuals, and instructions written or recorded on other media or devices such as, disk, tape, read-only memories, and computational methodologies, algorithms, and computer codes.

[30] The United States has an additional agreement with Australia for cooperation for the Separation of Isotopes by Laser Excitation (SILEX) technology for uranium enrichment.

[31] Pursuant to Section 6 of the Taiwan Relations Act (Title 22 U.S. Code section 3301 *et seq.*) and Executive Order 12143 (44 F.R. 37191), all agreements concluded with the Taiwan authorities prior to January 1, 1979, are administered on a nongovernmental basis by the American Institute in Taiwan, a nonprofit District of Columbia corporation, and constitute neither recognition of Taiwan authorities nor the continuation of any official relationship with Taiwan.

In: U.S. Nuclear Cooperation with Other Countries
Editor: Gabriel A. Oliveira

ISBN: 978-1-61209-910-1

Chapter 3

U.S.-AUSTRALIA CIVILIAN NUCLEAR COOPERATION: ISSUES FOR CONGRESS

Mary Beth Nikitin and Bruce Vaughn

SUMMARY

Australia and the United States have cooperated in the peaceful use of nuclear energy since the mid-1950s. The framework for this cooperation is a civilian nuclear cooperation agreement as required by section 123 of the Atomic Energy Act. President Obama transmitted the proposed text of the latest renewal agreement to Congress on May 5, 2010, along with the required Nuclear Proliferation Assessment Statement (NPAS) and his determination that the agreement promotes U.S. national security. Congress has 30 days of continuous session for consultations with the Administration, followed by an additional 60 days of continuous session to review the agreement. If not opposed by a joint resolution of disapproval or other legislation, then the agreement will be considered approved at the end of this time period. Congress also has the option of adopting either a joint resolution of approval with (or without) conditions or standalone legislation that could approve or disapprove the agreement. On November 30, 2010, the House passed H.R. 6411 by voice vote. The bill would approve the agreement even if the required congressional review period is not reached. The Senate has not yet acted on its version of the bill (S. 3844).

The United States and Australia first concluded a civilian nuclear cooperation agreement in 1957. That agreement was updated in 1979. Australia sells around 36% of its $1 billion in uranium exports to the United States. The United States is also a major processor of Australian uranium sold to other countries. Australia does not currently possess any nuclear power plants, but it operates one research reactor.

SUMMARY OF THE PROPOSED AGREEMENT

On May 5, 2010, President Obama transmitted the proposed text of a renewal of the U.S.-Australian civilian nuclear cooperation agreement to Congress for approval, along with the required Nuclear Proliferation Assessment (NPAS) and his determination that the agreement would promote U.S. national security. The annexed classified NPAS was to be submitted separately. The renewal agreement would replace the agreement in force from July 5, 1979, and expiring on January 16, 2011. The new agreement has an initial term of 30 years from the date of entry into force, and would continue in force for periods of five years each, unless terminated by either party.

According to President Obama's letter of transmittal, the agreement meets all the terms of the Atomic Energy Act[1] and therefore does not require any exemptions from the law's requirements. Therefore, the agreement will enter into effect after a 30-day consultation period and a review period of 60 days of continuous session,[2] unless Congress enacts a joint resolution of disapproval. This period began on May 5, 2010. Congress also has the option of adopting either a joint resolution of approval with (or without) conditions or standalone legislation that could approve or disapprove the agreement.

Section 123 of the U.S. Atomic Energy Act (AEA) of 1954 (42 U.S.C. 2011 et seq.) governs significant nuclear cooperation between the United States and other states.[3] The United States has agreements for civil nuclear cooperation in place with almost 50 countries. Such agreements, known as "123 agreements," provide the framework and authorization for cooperation, but do not guarantee certain exports, technology, or material. Before significant nuclear exports[4] can occur, the State Department, with the advice of the Department of Energy, negotiates an agreement, which must meet criteria listed in Section 123.a., (1) through (9), 42 U.S.C. 2151.[5]

The proposed agreement would permit the export, subject to licensing, of information, material, equipment, and components for nuclear research and nuclear power production. The agreement would not permit transfer of restricted data, sensitive nuclear facilities, or major critical components of those facilities. Sensitive nuclear technology may be transferred only if the agreement is amended. The proposed agreement does not allow for enrichment of uranium transferred under the agreement to levels of 20% U-235 or greater unless mutually agreed. The current agreement also does not allow for this. In accordance with the Nuclear Nonproliferation Act (NNPA) of 1978, the United States must approve of enrichment of any nuclear material exported from the United States to Australia. The proposed agreement modifies the 1979 agreement by expanding the scope of cooperation to cover nuclear forensics and allows the transfer of nuclear material for radioisotope production. It also updates the physical security requirements to reflect current international standards.

Issues for Congress

Uranium from Australia makes up 13% of U.S. uranium supplies. Renewal of the 123 agreement with Australia is necessary for continuing this cooperation. Having an agreement in place also facilitates Australia's export of uranium to countries where it will be used in a reactor with U.S. technology. Congress may consider Australia's role in developing multilateral nuclear fuel cycle services or assurances. Australia's strong bilateral relationship with the United States, political stability, and geographic location could be advantageous to the development of multilateral nuclear fuel facilities.

Enrichment technology developed in Australia has been transferred to the United States and is classified by both countries. In the event that Australia reverses current policy and decides to build a uranium enrichment plant on its territory, it would face several obstacles, and the United States would have to reconcile its policies. Uranium enrichment technology holders have committed to a ban on the export of enrichment technology until the Nuclear Suppliers Group decides on criteria for future transfers. It appears that Australia would qualify for transfer under this criteria, but enrichment plants would be a black box or turnkey facility. This type of transfer would be consistent with current U.S. policy. Australian officials have also noted additional concerns, such as cost and finding a local government willing to host a commercial enrichment plant.

As noted above, the Atomic Energy Act of 1954 provides for congressional review of proposed civilian nuclear cooperation agreements. The review comprises two periods of continuous session, one 30 days and one 60 days. The proposed agreement with Australia meets AEA requirements; therefore, it will enter into force after the review period is complete, absent congressional action. CRS estimates that the last day of the required congressional review period would be December 3, 2010 (this date is unofficial, and the sole definitive counters of the review periods are the House and Senate Parliamentarians). As the calendar stands as of mid-November, Congress is expected to adjourn on December 3, thus the required review period would be reached for this agreement and the agreement could enter into force without further action.

On June 15, 2010, House Foreign Affairs Committee Chairman Howard Berman and Ranking Member Ileana Ros-Lehtinen introduced H.J.Res. 88, which provides for the approval of the U.S.-Australia agreement. The House Foreign Affairs Committee held a hearing on September 24, 2010, about nuclear cooperation agreements, including the proposed renewal agreement with Australia.[6]

Once the congressional calendar for the remainder of 2010 became clearer in mid-September, some raised concerns that the 90-day period for congressional review of the U.S.-Australian renewal agreement would not be met. In this case, four scenarios are possible for approval for the agreement: (1) re-submittal of the agreement in the 112th Congress; (2) passage of a joint resolution of approval that would include language that waived the required 90-day review period; (3) separate legislation approving the proposed agreement with language to waive the 90- day requirement; (4) changes to the congressional calendar so that the lame duck session would contain enough days for the 90 day review period to lapse. Congress could also pass a joint resolution of disapproval or include its disapproval in separate legislation.

Senator Richard Lugar introduced a bill, "To provide for the approval of the Agreement Between the Government of the United States of America and the Government of Australia Concerning Peaceful Uses of Nuclear Energy" (S. 3844), on September 27, 2010. Representative Ileana RosLehtinen introduced the companion bill, H.R. 6411, on November 16, 2010. The House passed H.R. 6411 by voice vote on November 30, 2010.

It would approve the agreement notwithstanding the congressional review requirements of section 123 of the Atomic Energy Act. Other provisions of the AEA would apply. This bill would have the proposed agreement come into force on or after October 8, 2010.

Notwithstanding the provisions for congressional consideration of a proposed agreement for cooperation in subsection d. of section 123 of the Atomic Energy Act of 1954 (42 U.S.C. 2153), the Agreement Between the Government of the United States of America and the Government of Australia Concerning Peaceful Uses of Nuclear Energy, done at New York May 4, 2010, may become effective on or after October 8, 2010, as if all the requirements in such section 123 for consideration of such agreement had been satisfied, subject to subsection (b) of this section.

(b) Applicability of Atomic Energy Act of 1954 and Other Provisions of Law.—Upon coming into effect, the agreement referred to in subsection (a) shall be subject to the provisions of the Atomic Energy Act of 1954 (42 U.S.C. 2011 et seq.) and any other applicable United States law as if such agreement had come into effect in accordance with the requirements of section 123 of the Atomic Energy Act of 1954.

CONTEXT FOR DISCUSSION

The context for discussion of U.S.-Australian civilian nuclear cooperation is one set inside a very strong alliance relationship between the United States and Australia that stems from the two nations' shared democratic values, cultural and historic ties, and strategic outlook. According to U.S. Ambassador to Australia Jeffrey Bleich, "the relationship between the United States and Australia is one of our strongest and most productive international alliances. It is a bond solidified during the Pacific battles of World War II and strengthened over the ensuing decades."[7] The President's National Security Strategy document describes Australia, along with our other allies in Asia, as the "bedrock of security in Asia and a foundation of prosperity in the Asia-Pacific region."[8]

The Australia New Zealand United States (ANZUS) treaty was signed in 1951 and entered into force in 1 952.[9] A key aspect of the alliance with Australia today is the annual Australia-U.S. Ministerial (AUSMIN) meetings. These consultations have been held since 1985 and are attended by the Secretary of Defense and the Secretary of State from the United States, as well as the Australian Defense Minister and Minister for Foreign Affairs. The United States and Australia conduct joint combined military training exercises, have exchange officers in each others' armed forces, and work to standardize equipment and operational doctrine.[10] In recent years, Australia invoked the ANZUS Treaty after the attacks on the United States in 2001, and Australian troops fought with U.S. troops in Iraq and Afghanistan.

Australia has the largest non-NATO military contingent in Afghanistan[11] and fought alongside the United States in Vietnam, Korea, World War II, and World War I, as well as in Iraq and Afghanistan. Australia has also played a key role in promoting regional stability in places such as Timor Leste, the Solomon Islands, and Bougainville. The United States and Australia have a very close intelligence relationship. The United States and Australia also concluded a Free Trade Agreement, which came into force in 2005. The two nations signed a Defense Trade Cooperation Treaty in 2007 that would facilitate the trade of defense-related equipment and technology. The U.S. Senate has not yet voted on ratification of this treaty. For all of the above reasons, and others, the United States and Australia share a special relationship that includes a high degree of trust.

AUSTRALIAN NONPROLIFERATION POLICY

Australia also is a strong partner to the United States in nonproliferation policy. Australia has been a staunch supporter of the Non-Proliferation Treaty (NPT) since its adherence in 1973. Australia has promoted strengthened International Atomic Energy Agency (IAEA) safeguards on civilian nuclear activities and was the first to sign the Additional Protocol, which strengthened IAEA monitoring. Australia is also a member of the Nuclear Suppliers Group (NSG), an export control coordination body. [12]

Australia has developed stricter standards for nuclear trade than the export control regimes currently call for; for example, requiring an Additional Protocol to be in force for any country importing uranium from Australia. Over the past four decades, Australia has developed a bilateral safeguards system for uranium exports. A bilateral safeguards agreement is required, and countries may only retransfer Australian uranium to other countries that also have a bilateral safeguards agreement with Australia. [13]

In recent years, the international community has debated ways to foster the growth of nuclear energy while minimizing proliferation risks of the spread of sensitive nuclear technologies.[14] Australia has supported development of multilateral fuel assurance mechanisms while stating that a well-functioning market provides the best assurance of supply. At the 2010 NPT Review Conference, an Australian representative said that "Australia does not consider it necessary or desirable for each state with a nuclear program to develop national enrichment or reprocessing facilities ... it would be a perverse outcome from the standpoint of international security to see a proliferation of national fuel cycle facilities."[15] This is in line with current U.S. policy. The Nuclear Suppliers Group is currently debating criteria for future transfers of enrichment or reprocessing technologies.

Political division on nuclear power and to what extent Australia should pursue sensitive fuel cycle technologies was evident in earlier periods. Former Liberal Prime Minister John Gorton sought to keep Australia's nuclear options open, though it did not eventuate. He reportedly signed a secret deal with France to construct a uranium enrichment facility in Australia. A plan to develop a nuclear reactor at Jervis Bay was also considered. These plans were reversed by the election of Labor Party candidate Gough Whitlam, who was elected in 1972. Australia subsequently ratified the NPT in 1973. The Labor perspective on nuclear weapons was further promoted by Australian support for the South Pacific Nuclear Weapons Free Zone Treaty in 1985. Anti-nuclear sentiment in Australia was also influenced to an extent by the anti-nuclear movement in New Zealand, which led to New Zealand's de facto ouster from the ANZUS alliance in the mid-1980s and by the resumption of nuclear testing by France in French Polynesia in 1995.[16] Today, Australia remains a strong advocate of nuclear disarmament in international fora.

AUSTRALIA AND THE NUCLEAR BALANCE

While Australia is today a strong proponent of nuclear disarmament, arms control, and nonproliferation, this policy has evolved over time. As Rod Lyon has pointed out,[17] Australia's posture on nuclear weapons has been to a large extent determined by Australia's perceived "fit" in the regional and global strategic context. As the strategic environment changed from World War II to the era of the cold war to the present, Australia shifted from being a "possible nuclear proliferator, to a supporter of extended nuclear deterrence (which it benefits from under the ANZUS alliance) to an important advocate of global nuclear arms control."[18]

During the post-World War II period through the 1950s, Australia worked closely with its traditional security partner, the United Kingdom, including allowing Britain to conduct nuclear tests at an island off its northwest coast and in the desert.[19] Australian scientists worked in the British nuclear atomic bomb research program, and former Prime Minister Menzies reportedly saw such cooperation as an opportunity to gain knowledge of nuclear weapons development.[20] Australia received a nuclear research reactor from Britain in 1958.[21] It is apparently in the period after World War II and prior to closer Anglo-U.S. nuclear cooperation that Australia, as part of Britain's attempt to resurrect its empire, worked most closely with the United Kingdom in Britain's efforts to acquire its own nuclear deterrent to the Soviet Union.[22]

Australia's security environment became more precarious in the 1960s, which may have led it to consider its strategic options. Key events influencing this included China conducting its first nuclear test in 1964, the British announcement of its strategic departure from East of Suez in 1967, and the United States articulating the Guam Doctrine limiting U.S. ground troop support in 1969 while the war in Vietnam continued.

Australia's involvement in joint intelligence, communications, and early warning facilities with the United States made it a potential nuclear target of the Soviet Union during the cold war.[23] The Defense Support Program (DSP)[24] facility at Nurrungar, South Australia, was part of a constellation of geosynchronous satellites and ground tracking stations designed to detect the launch of ballistic missiles. Early warning of missile attack strengthened America's nuclear missile second strike capability, thereby strengthening nuclear deterrence. DSP later played a role in detecting Iraqi Scud missile launches during Operation Desert Storm. The Very Low Frequency (VLF) station involved in communications with American submarine deployments in the region at Northwest Cape, West Australia, and the signals intelligence collection facility at Pine Gap, Northern Territory, also played a key role in helping to maintain the bi-polar strategic balance of the cold war.

AUSTRALIA'S NUCLEAR CAPACITY

Although Australia is a major uranium exporter, it does not have any uranium enrichment plants, fuel fabrication facilities, or nuclear power plants. It currently operates a research reactor at Lucas Heights, used primarily for medical isotope production. The Australian Atomic Energy Commission (AAEC) was formed in 1952. The AAEC was replaced in 1987 by the Australian Nuclear Science and Technology Organization (ANSTO).[25] By statute, ANSTO cannot conduct any research in the design or production of nuclear weapons. Its main research facility, the Lucas Heights Research Establishment, opened in 1958. It was the site of Australia's first nuclear reactor, the High Flux Australian Reactor (HIFAR). Research and development work on uranium enrichment technology was also carried out at Lucas Heights (see below).

The nuclear fuel cycle begins with mining uranium ore and upgrading it to yellowcake. Because naturally occurring uranium lacks sufficient fissile ^{235}U to make fuel for commercial light-water reactors, the concentration of ^{235}U must be increased several times above its natural level of 0.7% in a uranium enrichment plant. A nuclear power plant operator or utility typically purchases yellowcake and contracts for its conversion to uranium hexafluoride, then enrichment, and finally fabrication into fuel elements. Australia exports its uranium after the mining and milling stage. Commercial enrichment services are available in the United States, Europe, Russia, and Japan. Fuel fabrication services are even more widely available. As noted above, Australia requires a bilateral safeguards agreement with uranium-importing countries to ensure its peaceful use.

R&D Activities

Research Reactors

Australia currently operates one research reactor. The Open Pool Australian Lightwater (OPAL) reactor was commissioned in 2007. It is a low-enriched uranium (LEU)-fueled 20-megawatt research reactor. It produces radioisotopes for medical purposes, among other research functions.[26] An older reactor, High Flux Australian Reactor (HIFAR), is in the process of being decommissioned (by 2018). Another research reactor, Moata, was decommissioned in 1995.

Uranium Enrichment

The AAEC conducted research into enrichment technologies starting in the 1960s. Work was done on both centrifuge and laser isotope separation enrichment. There had been discussions in the 1 970s about whether Australia should partner with Japan or others to build an enrichment plant in Australia for the purpose of exporting enriched uranium hexafluoride, which would be fabricated into fuel for power plants. These plans were highly controversial within Australia, were objected to by the United States, and were never completed.[27]

The Hawke government terminated uranium enrichment R&D by the AAEC in 1986 and shut down and dismantled Australia's laboratory-scale centrifuge enrichment plant. Australia continued to participate in the Hexapartite Safeguards Project, the purpose of which was to improve the IAEA's safeguards capability for enrichment plants. AAEC had also conducted research on molecular laser isotope separation (MLIS) enrichment, but discontinued its research.

Beginning in 1990, Silex Systems, Ltd. (Silex Systems) began R&D work on laser enrichment technology in a lab it leased at the Lucas Heights complex. The technology was proven on a laboratory scale in 1994 (known as SILEX, or Separation of Isotopes by Laser Excitation).[28] In 1996, the United States Enrichment Corporation (USEC, Inc.) signed an agreement for the development and licensing of SILEX technology for uranium enrichment. In 1999, the United States and Australia signed an agreement that allowed for the transfer of SILEX technology to the United States.[29] The SILEX technology was classified by both governments in 2001.

The U.S.-Australian SILEX agreement allows for technology export to United States but limits cooperation within the territory of Australia to research and development of SILEX technology. The agreement prohibits the construction of an enrichment facility with this technology in Australia unless the agreement is amended (Article 2.3). According to the Nuclear Proliferation Assessment Statement attached to the proposed agreement, Silex Systems stopped lab work on uranium enrichment in Australia but continues to support technology development in the United States. In 2006, GE/Hitachi Nuclear Energy signed an agreement with Silex Systems to license and commercialize laser separation technology for LEU production in the United States. GE/Hitachi's Global Laser Enrichment (GLE) subsidiary received a license from the Nuclear Regulatory Commission (NRC) for the building of a test facility in 2008. The "Test Loop" facility was built at GE's fuel fabrication facility in Wilmington, NC. GLE has applied for a commercial license from the NRC.[30]

Debate over whether Australia should have its own commercial uranium enrichment facility was reignited after Prime Minister Howard expressed support for the idea in 2006.[31] Howard argued that since Australia possesses the largest uranium reserves in the world, it would make economic sense to enrich uranium before exporting it, thereby adding to its value. John Carlson, director- general of the Australian Safeguards and Non-Proliferation Office, responded to press speculation at the time: "there are a number of uncertainties, including Australia's lack of an established enrichment technology and the license costs for imported technology, whether there would be investors prepared to outlay the substantial capital costs involved, and whether a State government would be willing to host an enrichment facility."[32] The present Australian government under the Labor Party appears to oppose building a uranium enrichment facility in Australia.

Uranium Mining and Milling

Australia has the world's largest reserves of uranium, with an estimated 23% or more of total reserves,[33] and is the world's third-largest exporter of uranium. The mining sector has become an increasingly important aspect of Australia's economy, accounting for 7% of GDP and 43% of exports. Mining is also thought of by many in Australia as one of the key reasons why Australia was able to avoid the recent global financial crisis.[34] As such, the Australian electorate may be in favor of exploring opportunities to expand mining exports, which could include the export of uranium to India. Australia provides 13% of U.S. uranium imports. More than 85% of Australia's uranium exports go to the United States, the European Union, and Japan.[35] Australia reached an agreement to sell uranium to China in 2006.[36] The Australian government is concluding arrangements for the export of uranium to Russia.

Uranium Sales to India

The Labor government of former Prime Minister Kevin Rudd opposed the export of uranium to India because India has not signed the NPT. Rudd was replaced by an intra-party change of leadership in June 2010 by Julia Gillard. She subsequently won a national election and remains prime minister. Her position on the issue is not yet clear. The John Howard government, in power until 2007, had supported uranium exports to India. The Nuclear Suppliers Group (NSG) decided in September 2008 to allow civilian nuclear exports to India under certain conditions. Australia did not actively block this consensus decision, but the Rudd government announced in September 2008 that Australia would not export uranium to

India, despite the NSG rule change. Australian Liberal Party opposition leader Tony Abbott has stated that he would reverse this policy and begin exports of uranium to India as part of an effort to expand Australia's ties to the region. Opinion polls conducted in May 2010 had Rudd and Abbott either even or with Abbott slightly ahead of Rudd.[37] The next Australian federal election will elect members of Parliament of Australia and must be held on or before April 2011. A Joint Feasibility Study recommended a bilateral free trade agreement with India in May 2010. India is Australia's fastest-growing export market.[38] Expanding trade and other ties with India could help Australia diversify its export markets, which have relied heavily on sales to China. Over the next 20 years, India will, according to some estimates, rapidly expand its middle class and embark on major infrastructure projects, which could fuel another resource boom in Australia.[39]

Nuclear Power Debate

The lead-up to the parliamentary elections of 2007 highlighted differences between the Kevin Rudd-led Labor Party and the Liberal-National Coalition led by former Prime Minister John Howard on nuclear energy policy. At the time, then Shadow Minister for Climate Change and the Environment Peter Garrett, formerly of the rock band Midnight Oil, warned that nuclear power was a bad idea for Australia, even given the demands of global warming, while Rudd declared nuclear power "not an option."[40] Bob Brown, the leader of the Green Party in Australia, is against nuclear power, as are many within the Labor Party.

Former Prime Minister John Howard stated in 2006 that he was of an "open mind" on nuclear power for Australia and that he believed that the "nuclear energy scene will change significantly in Australia."[41] The former Howard Liberal-National Coalition government selected Dr. Ziggy Switkowski to chair a uranium mining, processing, and nuclear energy review in 2006. The review's report viewed nuclear power as "a practical option for part of Australia's electricity production."[42] The report also noted that Australian demand for electricity will double by 2050 and that over this period, two-thirds of Australian power generation will need to be upgraded or replaced. While advocates of nuclear power remain within the Liberal-National Coalition, it appears that there was less support for the nuclear option within the coalition after Howard's defeat in 2007 than there was while he was prime minister.[43] The unpopularity of his position on nuclear power in the lead-up to the 2007 elections apparently forced Howard to modify his position.[44]

As Australia has become increasingly concerned about carbon emissions, a new consideration of nuclear power has emerged from some advocates. The low cost of Australian coal, relative to the cost of nuclear energy, appears to be a contributing factor as to why nuclear power has not been pursued. Dr. Switkowski has argued that a cost of carbon of $15 to $40 per ton would make nuclear power viable for Australia. Switkowski currently heads the Australian Nuclear Science and Technology Organization.[45] Australia has been considering a Carbon Pollution Reduction Scheme (CPRS), which in April 2010 the Rudd government shelved until at least 2013.[46] While the emissions trading scheme has been set aside for the moment, Australia continues to seek to reduce its carbon emissions.

End Notes

[1] Under section 123.a., codified at 42 U.S.C. 2153(a)), Atomic Energy Act of 1946, ch. 724, 60 Stat. 755 (1946), as amended. See also CRS Report RS22937, *Nuclear Cooperation with Other Countries: A Primer*, by Paul K. Kerr and Mary Beth Nikitin.

[2] Days on which either House is in a recess of more than three days (pursuant to a concurrent resolution authorizing the recess) do not count toward the total. If Congress adjourns its session *sine die*, continuity is broken and the count starts anew when it reconvenes.

[3] Nuclear cooperation includes the distribution of special nuclear material, source material, and byproduct material, to licensing for commercial, medical, and industrial purposes. These terms, "special nuclear material," "source material," and "byproduct material," as well as other terms used in the statute, are defined in 42 U.S.C. §20 14. See also CRS Report RS22937, *Nuclear Cooperation with Other Countries: A Primer*, by Paul K. Kerr and Mary Beth Nikitin.

[4] Significant nuclear cooperation includes the physical transfer of reactors, reactor components, or special nuclear material, source material, and byproduct material, under license for commercial, medical, and industrial purposes.

[5] The Atomic Energy Act also sets out procedures for licensing exports to states with whom the United States has nuclear cooperation agreements. (Sections 126, 127, and 128 codified as amended at 42 U.S.C. 2155, 2156, 2157.) Even with a 123 agreement in place, each export of nuclear material, equipment, or technology requires a specific export license or other authorization.

[6] House Foreign Affairs Committee Hearing, "Nuclear Cooperation After Iran and Khan: Are We Asking Enough of Current and Future Agreements?" September 24, 2010.

[7] Jeffrey Bleich, "Statement Before the Senate Foreign Relations Committee," Washington, DC, November 5, 2009.

[8] "National Security Strategy, The White House, Washington, DC, May 2010.

[9] For a text of the ANZUS Treaty, see http://avalon.law.yale.edu/20th_century/usmu002.asp.

[10] "Background Note: Australia," the Department of State, November 23, 2009.

[11] Australian Government, Department of Defence Media Release, "Republic of Korea Joins ISAF as a Non NATO Contributing Nation," April 22, 2010.

[12] See also CRS Report RL3 1559, *Proliferation Control Regimes: Background and Status*, coordinated by Mary Beth Nikitin.

[13] "Australia's Network of Nuclear Safeguards Agreements," Australian Department of Foreign Affairs and Trade, http://www.dfat.gov.au/security

[14] CRS Report RL34234, Managing the Nuclear Fuel Cycle: Policy Implications of Expanding Global Access to Nuclear Power, coordinated by Mary Beth Nikitin.

[15] Statement by Mr. Jeremy Kruse to Main Committee III regarding the IAEA Peaceful Uses Initiative, as part of the 2010 Review Conference of the Parties to the Treaty on the Non-proliferation of Nuclear Weapons, as delivered on May 10, 2010.

[16] Jeffrey Lantis, "Elections and Enduring Realities: Australia's Nuclear Debate," *Arms Control Today*, April 2008.

[17] Rod Lyon, "Australia: Back to the Future," in Muthiah Alagappa ed. *The Long Shadow: Nuclear Weapons and Security in 21st Century Asia* (Stanford: Stanford University Press, 2008).

[18] Rod Lyon, "Australia: Back to the Future," in Muthiah Alagappa ed. *The Long Shadow: Nuclear Weapons and Security in 21st Century Asia* (Stanford: Stanford University Press, 2008).

[19] Lorna Arnold, *A Very Special Relationship: British Atomic Weapon Trials in Australia* (London: Her Majesty's Stationery Office, 1987).

[20] Jeffrey Lantis, "Elections and Enduring Realities: Australia's Nuclear Debate," *Arms Control Today*, April 2008.

[21] Jeffrey Lantis, "Elections and Enduring Realities: Australia's Nuclear Debate," *Arms Control Today*, April 2008.

[22] Wayne Reynolds, *Australia's Bid for the Atomic Bomb* (Melbourne: Melbourne University Press, 2000).

[23] Cameron Stewart et al., "US Aussie Spy Base Revelations," *The Australian*, February 18, 1999.

[24] Jeffrey Richelson, *America's Space Sentinels*, University of Kansas Press (1999).

[25] http://www.ansto.gov.au/discovering_ansto/history

[26] http://www.ansto.gov.au/discovering_ansto/anstos_research_reactor.

[27] Wayne Reynolds, "Australia's quest to enrich uranium and the Whitlam Government's loans affair," *The Australian Journal of Politics and History*, vol. 54, no. 4 (December 2008).

[28] "Company History" page at http://www.silex.com.au/.

[29] Agreement for Cooperation Between Australia and the United States of America Concerning Technology for the Separation of Isotopes of Uranium by Laser Excitation.

[30] The NRC's estimated 30-month review period began in August 2009. Silex Annual Report 2009, http://www.silex.com.au/SilexAR2009.pdf.

[31] Raymond Bonner, "Call to Enrich Uranium in Australia Stirs Debate," *The New York Times*, August 2, 2006; "Uranium enrichment program revived after 20 years," *Australian Broadcasting Corporation*, June 14, 2007.

[32] "The Prospects for Uranium Enrichment in Australia—Correcting the Record," Statement by Mr. John Carlson, Director-General, Australian Safeguards and Non-proliferation Office, June 18, 2007, http://www.dfat.gov.au/media releases/department/d010_07.html.
[33] "Australia's Uranium," January 2010, http://www.world-nuclear.org/info/inf48.html.
[34] "Heroes in Hard Hats," The Sydney Morning Herald, May 22, 2010.
[35] Unclassified Nuclear Proliferation Assessment Statement Pursuant to Section 123 a. of the Atomic Energy Act of 1954, as Amended with Respect to the Proposed Agreement between the Government of the United States of American and the Government of Australia Concerning Peaceful Uses of Nuclear Energy, May 4, 2010.
[36] "Australia, China Sign uranium Deal," *CNN,* April 3, 2006.
[37] "Australia's Abbott Wants Japan Trade Deal, India Uranium Sale," *Bloomsburg Business Week*, May 26, 2010.
[38] "Free Trade Pact with India More Likely," *Australian Associated Press,* May 5, 2010.
[39] Robin Bromby, "Massive Indian Expansion to Fuel Ongoing Resource Demand," The Australian, May 17, 2010.
[40] Jeffrey Lantis, "Elections and Enduring Realities: Australia's Nuclear Debate," *Arms Control Today*, April 2008.
[41] Michael Perry, "Australia May Develop Nuclear Power—Howard," *Ezilon.com* May 20, 2006.
[42] Department of the Prime Minister and Cabinet, *Uranium Mining, Processing and Nuclear Energ—Opportunities for Australia?* Report to the Prime Minister by the Uranium Mining, Processing and Nuclear Energy Review Task Force, Commonwealth of Australia, 2006.
[43] Steve Lewis, "MacFarlane Says Australia Must Adopt Nuclear Power," *Courier Mail*, August 19, 2008.
[44] Jewel Topsfield, "Howard's Nuclear Meltdown," *The Age*, August 24, 2007.
[45] Australian Government, Australian Nuclear Science and Technology Office, http://www.ansto.gov.au.
[46] "Australia Cans its Carbon Emissions Trading Scheme for Now," *East Asia Forum*, May 13, 2010.

In: U.S. Nuclear Cooperation with Other Countries
Editor: Gabriel A. Oliveira

ISBN: 978-1-61209-910-1

Chapter 4

U.S.-RUSSIAN CIVILIAN NUCLEAR COOPERATION AGREEMENT: ISSUES FOR CONGRESS

Mary Beth Nikitin

SUMMARY

The United States and Russia signed a civilian nuclear cooperation agreement on May 6, 2008. President Bush submitted the agreement to Congress on May 13. The agreement was withdrawn from congressional consideration by President George W. Bush on September 8, 2008, in response to Russia's military actions in Georgia. President Obama transmitted the proposed text of the agreement to Congress on May 10, 2010, along with the required Nuclear Proliferation Assessment (NPAS) and his determination that the agreement promotes U.S. national security. Congress has 30 days of continuous session for consultations with the Administration, followed by an additional 60 days of continuous session to review the agreement. If not opposed by a joint resolution of disapproval or other legislation, then the agreement will be considered approved at the end of this time period.

This report discusses key policy issues related to the agreement, including future nuclear energy cooperation with Russia, U.S.-Russian bilateral relations, nonproliferation cooperation, and Russian policies toward Iran. These issues were also relevant to the debate when the agreement was being considered in the 110th Congress.

This report will be updated as events warrant.

On May 10, 2010, President Obama transmitted the proposed text of the U.S.-Russian civilian nuclear cooperation agreement to Congress for approval, along with the required Nuclear Proliferation Assessment (NPAS) and his determination that the agreement promotes U.S. national security. The annexed classified NPAS was to be submitted separately. The agreement was signed by the two countries in Moscow on May 6, 2008. President George W. Bush first submitted it to Congress on May 13, 2008,[1] but in September 2008 rescinded the national security determination following Russian military actions in the Republic of Georgia. This had the effect of removing the agreement from congressional consideration. President Obama stated his commitment to seeing the agreement enter into force in summit statements with Russian President Medvedev in April and July 2009. President Obama's May 10, 2010, letter of transmittal says that the situation in Georgia is no longer an obstacle and that "the level and scope of U.S.-Russian cooperation on Iran are sufficient to justify resubmitting the proposed agreement."[2]

According to President Obama's letter, the agreement meets all the terms of the Atomic Energy Act[3] and therefore does not require any exemptions from the law's requirements. Therefore, the agreement will enter into effect after a 30-day consultation period and a review period of 60 days of continuous session[4] unless Congress enacts a joint resolution of disapproval. Congress also has the option of adopting either a joint resolution of approval with (or without) conditions, or standalone legislation that could approve or disapprove the agreement.[5]

The agreement would permit the export, subject to licensing, of technology, material, equipment, and components for nuclear research and nuclear power production. The agreement would not permit transfer of restricted data. The agreement need to be amended before any transfer of sensitive nuclear technology, sensitive nuclear facilities, and major critical components of those facilities. The parties would also need to agree to reprocessing of material transferred under the agreement. Some limited enrichment and blending or down-blending for LEU fuel production would be permitted.

Background

Section 123 of the U.S. Atomic Energy Act (AEA) of 1954 (42 U.S.C. 2011 et seq.) governs significant nuclear cooperation between the United States and other states.[6] The United States has agreements for civil nuclear cooperation in place with almost 50 countries. Such agreements, known as "123 agreements," provide the framework and authorization for cooperation, but do not guarantee certain exports, technology, or material. Before significant nuclear exports[7] can occur, the State Department, with the advice of the Department of Energy, negotiates an agreement, which must meet criteria listed in Section 123.a., (1) through (9), 42 U.S.C. 2151.[8] Russia is the only NPT-recognized nuclear weapon state and the only major nuclear energy producer with whom the United States does not yet have a civilian nuclear cooperation agreement. In the case of the agreement with China, Members of Congress attached conditions to a joint resolution of approval of the agreement, based on concerns, among others, that China was exporting materials and equipment relevant for nuclear weapons development to non-nuclear weapon states.[9]

Cooperation between the United States and Russia on civilian nuclear energy is not new, but the level has fluctuated depending on broader political developments. The United States and the Soviet Union concluded a limited 10-year agreement for nuclear cooperation in 1973 to allow for cooperation in controlled thermonuclear fusion, fast breeder reactors, and fundamental research. The 1973 agreement also established a Joint Committee on Cooperation in the Peaceful Uses of Atomic Energy that was to meet annually. This agreement was extended in 1983 and in 1988, and exchanges on safety practices significantly increased after the 1986 Chernobyl power plant accident. The two superpowers convened a Joint Coordinating Committee for Civilian Reactor Safety starting in 1988.[10]

After the fall of the Soviet Union and prior to July 2006, Moscow's nuclear commerce with Iran presented the chief obstacle to concluding a broad U.S.-Russian nuclear cooperation under section 123. Project-specific agreements were concluded throughout this period. Several factors may have contributed to the shift in U.S. policy under the George W. Bush Administration: a tougher line by Moscow since 2003 with respect to Iran, especially Russia's agreement with Iran to take back spent nuclear fuel from the Russian-built Bushehr reactor; President Bush's embrace of nuclear power as an alternative to reliance on hydrocarbons; President Bush's proposals to multi-lateralize the nuclear fuel cycle and develop proliferation-resistant technologies through the Global Nuclear Energy Partnership (GNEP); and Russia's own proposals to host an international fuel center that would store and reprocess spent fuel and enrich uranium for fresh fuel.

Under the Obama Administration, officials have expressed support for the nuclear cooperation agreement with Russia, but were waiting for the "appropriate time" to submit the agreement to Congress.[11] President Obama's letter of May 10, 2010, outlines ways in which the current Administration sees this agreement as being beneficial for U.S. interests, primarily in that it would contribute to "the growth of clean, safe and secure nuclear energy for peaceful purposes."

The letter cites several areas of recent "progress" in cooperation between the United States and Russia:

- Russian support for a new United Nations Security Council Resolution on Iran.
- Signature on April 8, 2010, of the New START Treaty that would reduce the number of deployed strategic nuclear weapons.[12]
- Signature on April 13, 2010, of the Protocol to amend the 2000 U.S. Russian Plutonium Management and Disposition Agreement, which will dispose of at least 34 metric tons of excess weapons plutonium in each country.
- Russia's establishment of an international nuclear fuel reserve in Angarsk.
- Continued joint support for the Global Initiative to Combat Nuclear Terrorism.

Congressional Consideration

Congressional debate over the agreement in the past has focused on several key issues: the nature of Russian-Iranian cooperation, the impact of a U.S.-Russian agreement on the future of nuclear fuel cycle policies, and the impact of the agreement on bilateral relations including nuclear nonproliferation cooperation. While some view the agreement as promoting bilateral cooperation on U.S. nonproliferation goals and as a recognition of recent Russian cooperation, others believe the United States could gain leverage on negotiations with Russia on Iran and other matters by delaying approval of the agreement.

Legislation in the 110th Congress

Even before the crisis in the Republic of Georgia in August 2008, approval of a U.S.-Russia 123 agreement by Congress was not certain. Legislation both supporting and opposing the agreement was introduced in the 110th Congress:

- Representative Edward Markey on May 14, 2008, introduced H.J.Res. 85 expressing disfavor of the agreement.
- On June 24, 2008, Chairman of the Senate Foreign Relations Committee Joseph Biden and Senator Richard Lugar submitted a joint resolution of approval, S.J.Res. 42. It was discharged from committee but indefinitely postponed by unanimous consent in September 2008.
- Chairman of the House Committee on Foreign Affairs Howard Berman and Ranking Member Ileana Ros-Lehtinen introduced a resolution of disapproval, H.J.Res. 95, on June 24, 2008.
- The House Committee on Foreign Affairs reported H.R. 6574[13] on July 23, 2008. This bill would have approved the U.S.-Russia 123 agreement, notwithstanding the AEA, with certain conditions. Under this resolution, no license could be issued for the export of nuclear material, equipment, or technology to Russia unless the President certified to Congress that Russia (1) is not transferring sensitive nuclear, biological- or chemical-weapons-related, ballistic or cruise missile technologies, goods, or services to Iran; (2) is cooperating with the United States on international sanctions on Iran; and (3) had ratified appropriate nuclear liability conventions or enacted domestic laws to protect U.S. firms.

In response to Russian actions in August over the conflict in Georgia, some Members of Congress called on the Bush Administration to withdraw the agreement from congressional consideration.[14]

There was no precedent for the President withdrawing a 123 from congressional consideration, and the Atomic Energy Act does not specify procedures for doing so. On September 8, 2008, the Secretary of State issued a statement saying that the President would notify Congress that "he has today rescinded his prior determination" regarding the agreement and therefore there is no basis for Congress to consider it. Secretary Rice states that "the U.S. nonproliferation goals contained in the proposed Agreement remain valid: to provide a sound basis for U.S.-Russian civil nuclear cooperation, create commercial opportunities, and enhance cooperation with Russia on important global nonproliferation issues." She expresses regret for the decision but says that "unfortunately, given the current environment, the time is not right for this agreement."[15] In his message to Congress, President Bush wrote that this decision is "in view of recent actions by the Government of the Russian Federation incompatible with peaceful relations with its sovereign and democratic neighbor Georgia." In the original determination of May 5, 2008 (Presidential Determination 2008-19), the President determined that the agreement will promote and will not pose an unreasonable risk "to the common defense and security."[16] The President's message of September 8 says this determination "is no longer effective." It also says that "if circumstances should permit future reconsideration of the proposed Agreement, a new determination will be made and the proposed Agreement will be submitted for congressional review pursuant to section 123 of the Act."[17]

Additional legislation proposed in the 110th Congress focused on Iran's nuclear programs and also sought to condition nuclear cooperation with Russia. The Iran Counter-Proliferation Act of 2007 (H.R. 1400), passed by the House, would prohibit any "agreement for cooperation between the United States and the government of any country that is assisting the nuclear program of Iran or transferring advanced conventional weapons or missiles to Iran."

Similarly, S. 970 specifically would have prohibited a 123 agreement with Russia until "Russia has suspended all nuclear assistance to Iran and all transfers of advanced conventional weapons and missiles to Iran" or "Iran has completely, verifiably, and irreversibly dismantled all nuclear enrichment-related and reprocessing-related programs."

The Iran Sanctions Act of 2008 (S. 3227) included a prohibition on entering into a nuclear cooperation agreement with Russia or granting licenses for the direct or indirect export or the direct or indirect transfer of nuclear-related goods, services, or technologies to Russia until certain presidential certifications are made. S. 3227 was reported out of the Senate Finance Committee on July 7, 2008, but was not passed by the full Senate.

The Security through Termination of Proliferation Act of 2008 (H.R. 6178, introduced on June 4, 2008) included similar provisions, including that a nuclear cooperation agreement with a country proliferating to Iran, North Korea, or Syria may not enter into force. These bills, as well as letters sent to the President from Members of Congress after submittal of the 123 agreement to the Congress, showed a linkage between Russia's policies toward Iran and support for a bilateral civilian nuclear accord in Congress.

The 2008 Nuclear Proliferation Assessment Statement (NPAS)

In 2008, some Members of Congress raised concerns about the information contained in the Nuclear Proliferation Assessment Statement (NPAS). The NPAS is described in section 123.a. (42 U.S.C. 2153(a)) and is required part of a 123 agreement package for Congress. An unclassified NPAS is submitted along with the proposed text of the agreement, and a classified annex is submitted separately. The NPAS is to be prepared by the State Department in consultation with the Director of National Intelligence. Its purpose is to explain how the agreement meets the AEA nonproliferation requirements. The unclassified NPAS usually includes an overview of the country's nuclear energy program and related infrastructure, nuclear weapons program (if relevant), nonproliferation policies, and relations with third countries of concern in the nuclear arena.

Some Members of Congress were concerned that the 2008 NPAS for the US-Russia 123 agreement excluded information regarding nuclear trade between Russia and Iran. This prompted then-Chairman of the House Energy and Commerce Committee John Dingell and Subcommittee on Oversight and Investigations Chairman Bart Stupak to request that the Government Accountability Office (GAO) evaluate the inter-agency process for development of U.S.-Russia NPAS, whether the NPAS conclusions were supported, and whether any information was omitted that might change these conclusions.[18] The GAO also lists Chairman Henry Waxman and Representative Edward Markey as report requesters. The GAO issued its report in July 2009.[19] The findings were related primarily to the inter-agency review process and recommended that the State Department should clarify interagency roles, allow adequate time for review by the intelligence community and the Nuclear Regulatory Commission, and establish written procedures for development and review of 123 agreements and associated documents.

Legislation in the 111th Congress

Upon the Obama Administration's transmittal of the agreement to Congress in 2010, Chairman of the House Foreign Affairs Committee Howard Berman said that the top nonproliferation policy priority should be preventing Iran from obtaining nuclear weapons and that "at the appropriate time, we will examine the agreement more fully."[20] Three joint resolutions were introduced in the House, and referred to the House Foreign Affairs Committee, two expressing disfavor and one providing for approval of the agreement:

- On May 21, 2010, Representative Edward Markey and co-sponsor Representative Jeff Fortenberry introduced a joint resolution (H.J.Res. 85) expressing disfavor of the proposed agreement.
- On June 21, 2010, House Foreign Affairs Committee Chairman Howard Berman and co-sponsor Representative Dana Rohrabacher introduced a joint resolution that provides for approval of the proposed agreement (H.J.Res. 91).
- On June 21, 2010, House Foreign Affairs Committee Ranking Member Ileana Ros-Lehtinen, with Representatives Dan Burton and Edward Royce, introduced a joint resolution that provides for disapproval of the proposed agreement (H.J.Res. 92).

Senate Foreign Relations Committee Chairman John Kerry and Ranking Member Richard Lugar introduced a joint resolution (S.J.Res. 34) that would approve the proposed agreement on June 21, 2010. It was referred to the Senate Foreign Relations Committee. As stated above, no positive action by Congress is required for the agreement to enter into force after the congressional review period expires.

The Comprehensive Iran Sanctions, Accountability, and Divestment Act of 2010

The Comprehensive Iran Sanctions, Accountability, and Divestment Act of 2010 was signed by the President on July 1, 2010 (P.L. 111-195). Section 3 (9) of the bill says that it is the sense of Congress that no export licenses should be given under a civilian nuclear cooperation (123) agreement if the recipient country "is providing similar nuclear material, facilities, components, or other goods, services, or technology to another country that is not in full compliance with its obligations under the Nuclear Non-Proliferation Treaty, including its obligations under the safeguards agreement between that country and the International Atomic Energy Agency," unless the President determines that such transfers would not undermine U.S. nonproliferation objectives.

Section 102, subsection a, of the bill prohibits the issuance of export licenses under a 123 agreement for "any country whose nationals have engaged in activities with Iran relating to the acquisition or development of nuclear weapons or related technology, or of missiles or other advanced conventional weapons that have been designed or modified to deliver a nuclear weapon." The President can waive the provision by making a determination and notification to the appropriate congressional committees that the country did not know or have reason to know about the activity, or if the country is taking "all reasonable steps" to prevent recurrence and penalize the person involved.

An earlier House report (H.Rept. 111-342) states that "the Committee believes that a country that is, as a matter of policy or through willful inaction, allowing its citizens or companies to provide equipment, technology or materials to Iran that make a material contribution to its nuclear capabilities should not at the same time benefit from nuclear cooperation with the United States."

NUCLEAR ENERGY AND NONPROLIFERATION COOPERATION

The United States and Russia currently cooperate on a variety of nuclear nonproliferation and nuclear energy initiatives, under ad hoc agreements.[21] While this cooperation is focused primarily on nuclear nonproliferation measures, in recent years the United States and Russia have explored ways to develop civilian nuclear energy cooperation. Presidents Bush and Putin in July 2006 established a working group[22] whose report defined an Action Plan for cooperation that led to the bilateral Presidential Declaration on Nuclear Energy and Nonproliferation of July 3, 2007.[23] In an effort to continue this process and as part of the Obama Administration's "reset" of relations with Russia, in July 2009 Presidents Obama and Medvedev established a Bilateral Presidential Commission that included a Working Group on Nuclear Energy and Security chaired by Sergei V. Kiriyenko, Head of Rosatom, and Daniel Poneman, Deputy Secretary of Energy. The July 2009 Joint Statement reaffirmed their intention to bring a bilateral nuclear cooperation agreement into force and said that the two countries would focus on:

- development of prospective and innovative nuclear energy systems;
- research into methods and mechanisms for the provision of reliable nuclear fuel cycle services;
- research into international approaches for the establishment of nuclear fuel cycle services to secure the nuclear weapons nonproliferation regime; and
- improvement of the international safeguards system.

The working group agreed on an Action Plan for nuclear security and civil nuclear energy cooperation in September 2009. A commission report said that the working group identified research and development areas for collaboration and is working on a "new fuel services framework."[24] An argument in favor of the agreement is that the United States could gain from Russian advanced fuel cycle research and development experience.[25] Because the United States does not operate fast neutron reactors or reprocess, testing of fuels could be done in Russia, including post-irradiation examination. Supporters argue that U.S. partnership in developing these technologies could help ensure that proliferation resistance remains a priority. On the other hand, critics point out that the agreement risks entrenching a policy of accepting reprocessing as a necessary part of the future of nuclear energy and that this would raise proliferation risks.

The proposed agreement could provide Russia with access to U.S. nuclear technologies and markets, and would open the possibility of receiving U.S.-origin nuclear materials into Russia for storage or processing. Some argue that the agreement might be construed as a U.S. stamp of approval for Russia's civilian nuclear industry when safety and environmental problems with the Russian nuclear industry remain. Others counter that only through such an agreement will U.S. safety technology and standards be available to Russia. Russia could potentially expand its reach into new nuclear power markets by adding U.S. safety and automated control systems to its exported reactors, or partnering with U.S. multinationals. Russia could also potentially improve the operational efficiency of its own reactors with U.S. technology.[26]

The United States and Russia both promote a future global nuclear energy framework that addresses emerging nuclear energy states' fuel needs while dissuading them from pursuing indigenous enrichment and reprocessing technologies. This includes a "cradle to grave" approach to nuclear fuel. As part of this effort, recent Russian nuclear power plant exports, such as with Turkey, are a "package deal" that would include construction, operation, fuel services, and spent fuel return.[27] Broader proposals to discourage new states from building fuel cycle facilities include the development of multilateral fuel assurances and international fuel service centers.[28] For this purpose, Russia has set up the joint venture International Uranium Enrichment Center at Angarsk, which is under international safeguards. An international LEU fuel reserve will also be hosted at the site. Proponents of the agreement say that collaboration between the United States and Russia on providing nuclear fuel cycle services to nonnuclear weapon states could increase the confidence of customer states and therefore increase participation.

Experts and policy makers have also been exploring what role Russia could play in addressing the issue of nuclear waste and spent fuel disposition. Some have proposed that a 123 agreement with Russia could open the possibility of reprocessing U.S.-origin spent fuel from third countries (although Russia has not yet decided to do this) or long-term spent fuel storage of such material in Russia.[29] The enrichment of U.S.-obligated reprocessed uranium, and the re-enrichment of U.S. uranium tails or U.S.-origin tails, using Russian enrichment facilities, could likewise occur only if a 123 agreement was in force. [30] Under Article 9 of the proposed agreement, conversion and enrichment to less than 20%, fabrication of LEU fuel, irradiation, blending, or down-blending to LEU would be permitted under the agreement. The parties would have to agree to reprocessing of U.S.-origin spent fuel before this occurred.

Nuclear Liability

For these potential areas of cooperation to be realized, however, nuclear liability coverage for U.S. companies doing business in Russia would need to be clarified. The Russian Federation has been party to the Vienna Convention on Civil Liability for Nuclear Damage since 2005. However, Rosatom Corporation enjoys sovereign immunity as a partially state-owned enterprise. Russia has not yet signed or ratified the Convention on Supplementary Compensation for Nuclear Damage (CSC).[31] Currently, ad hoc bilateral agreements with liability coverage are in place for U.S. safety and nonproliferation assistance programs. Some U.S. companies have stated that they would need Russia to ratify the CSC or adopt domestic laws that would provide liability protection for U.S. firms before doing business in Russia. In a 2003 letter to the Bush Administration, the Contractors International Group on Nuclear Liability (CIGNL) wrote that "the various bilateral and multilateral indemnity agreements that have been concluded to date are not considered to provide adequate nuclear liability protection by most large, well-capitalized U.S. companies."[32] As cited above, the proposed legislation reported out of the House Foreign Affairs Committee (H.R. 6574) in 2008 would have approved the agreement with conditions that included the stipulation that Russia would have to ratify appropriate nuclear liability conventions or enact domestic laws to protect U.S. firms before a license under the agreement could be issued.

U.S.-Russian Relations[33]

The main focus of U.S.-Russia relations over the past year has been the negotiation of a follow-on Strategic Arms Reduction Treaty.[34] However, Presidents Obama and Medvedev set up a process to review engagement in many sectors, as part of a "reset." The NATO-Russia Council resumed its meetings in April 2009, and in July 2009, the Russian president announced that Russia would grant supply rights to NATO forces in Afghanistan overland and in Russian airspace. Differences remain over a number of foreign policy issues, particularly Russian military bases in and diplomatic recognition of Abkhazia and South Ossetia, characteristics of future missile defense systems in Eastern Europe, the expansion of NATO and how to deal with the Iranian nuclear program. In this context, some argue that expanding cooperation with Russia on civilian nuclear energy is premature. Others argue that nonproliferation, nuclear terrorism prevention, and nuclear energy may have particular value for the bilateral relationship in this context, and that a 123 agreement could be used to influence Russian policies.[35]

U.S. Ambassador Burns remarked at the May 2008 signing ceremony that the 123 agreement marks Washington and Moscow's transition from "nuclear rivals" to "nuclear partners." Although a 123 agreement will not itself stipulate new programs or collaborative projects, it may have symbolic value and remove a longtime irritant in bilateral relations. Supporters argue that rejecting the agreement could embolden anti-U.S. sentiment and be counter-productive to cooperation in other areas. Critics counter that its symbolic value is a reason not to enact it at this time—it could be viewed as a reward for a Russian government that critics view as antidemocratic and repressive, and whose foreign policy often has been at odds with U.S. interests. President Bush's rescission of the national security determination as related to the proposed 123 agreement in 2008 following Russian military actions in Georgia clearly demonstrated the possibility of other foreign policy priorities overshadowing U.S.-Russian nuclear energy cooperation.

Russian Policy toward Iran

During the Clinton Administration and the early Bush Administration, the United States had a policy not to conclude a civilian nuclear cooperation agreement with Russia while it was building a nuclear power reactor for Iran at Bushehr. After details about Iran's clandestine nuclear activities came to light during 2002-2006, Russia began to step up cooperation with the United States and other countries negotiating with Iran over its nuclear program. Russia has insisted on IAEA safeguards on any transfers to Iran's civilian nuclear reactor at Bushehr. The 2005 Russian- Iranian agreement on fuel supply for Bushehr requires the return of spent fuel to Russia, in order to prevent Iran from extracting plutonium from the spent fuel. Moscow also invited Tehran to participate in its newly established international uranium enrichment center at Angarsk, as an alternative to an indigenous Iranian enrichment capability—an offer that Iran has rejected. The Bush Administration supported this approach and since 2002 no longer objected to Russia's building the Bushehr nuclear power plant in Iran. The Bush and Obama Administrations viewed the Russian provision and take-back of nuclear fuel for the Bushehr reactor as demonstrating that it is possible for Iran to have access to nuclear energy without developing its own fuel cycle. [36]

Russia has generally been only reluctantly supportive of U.N. Security Council Resolutions (UNSCRs) imposing sanctions on Iran, preferring a primarily diplomatic solution to the crisis. However, President Putin signed decrees to fully implement UNSCRs 1737, 1747, and 1803 in 2008,[37] and Russia also supported UNSCR 1835. In 2009, Russia and the United States worked closely on proposals to supply the Tehran Research Reactor with fuel.[38] In May 2010, Russia announced its support for a new U.N. Security Council sanctions resolution, an objective of the Obama Administration. In general, analysts argue that Russian and American views about the nature of the Iranian nuclear program have come closer in recent years, particularly following the revelation in September 2009 of the enrichment facility being built at Qom.

Continued questions about the nature and extent of Russian cooperation with Iran remain a potential obstacle to future approval of the 123 agreement by Congress. The 2006 Iran Freedom Support Act (P.L. 109-293) stated the sense of Congress that no nuclear cooperation agreement should be entered into with a country that is assisting the nuclear program of Iran. As noted above, the Comprehensive Iran Sanctions, Accountability, and Divestment Act of 2009 (H.R. 2194) amended the Iran Support Act to prohibit peaceful nuclear cooperation with a country if one of its citizens or companies was assisting Iran in its nuclear weapons program.

Both the 2008 and 2010 NPAS state that the United States "has received assurances from Russia at the highest levels that its government would not tolerate cooperation with Iran in violation of its U.N. Security Council obligations." Some reports in 2008 said that Russian entities had transferred sensitive nuclear technology to Iran, but this activity was ended by high-level Russian governmental intervention and assurances were given to the highest levels of the U.S. government.[39] Additional details on the proliferation concerns associated with Russian-Iranian cooperation are likely part of the classified annex.

The 2009 Director of National Intelligence report to Congress on WMD Acquisition says that "entities in Russia and China continue to sell technologies and components in the Middle East and South Asia that are dual use and could support WMD and missile programs.... Russian entities have provided assistance to missile and civil nuclear programs in Iran and India." The report also says that Russian entities have been a source of dual-use biotechnology equipment and expertise, including for Iran.[40]

Another issue of congressional interest is the planned Russian sale of five S-300 air defense systems to Iran. Russia has so far stalled on completing this transaction. It is expected to be deployed near Iranian nuclear facilities. The May 2010 draft of the U.N. Security Council sanctions resolution on Iran would not prohibit this transfer. Air defense systems are not prohibited under international export control regimes, nor would the transfer automatically fall under any current U.S. sanctions.

On May 21, 2010, the State Department announced it was removing sanctions on four sanctioned entities in Russia.[41] Since 1998, at least 19 different Russian entities have been placed under proliferation-related sanctions on over 20 different occasions. However, with the removals on May 21, there appear to be no current proliferation-related sanctions against Russian entities.[42] Some observers have asserted that removal of sanctions was done in exchange for Russian support for a draft U.N. Security Council resolution that strengthens sanctions against Iran. The State Department spokesman has said that, regardless, there was no evidence that the companies removed from the sanctions list were currently transferring arms or technology.[43] In March 2010, the Administration lifted sanctions on two other Russian entities, Glavkosmos and the Baltic State Technical University, both sanctioned in 1998 for helping Iran's missile and weapons programs.

Additionally, the Department of Commerce lists entities subject to license requirements for proliferation-related end-use or end-user controls under Part 744 Supplement of the Export Administration Regulations (EAR). As of February 19, 2010, there were eight Russian entities on this list.[44] Three of these entities' license applications are reviewed on a case-by-case basis, while five are under "presumption of denial." According to Commerce Department officials, this list is currently under review.

Some argue that maximum leverage has already been gained in coaxing Russian behavior on Iran in exchange for the signing of a 123 agreement, and that there will be opportunities in the future to exercise further leverage if necessary, because each transaction under a 123 agreement must be approved for licensing.[45] Supporters may also see the 123 agreement as a way to encourage Russia to continue pressing Iran on such issues as the Bushehr reactor safeguards. Some argue that engaging Russia on the scientific level would improve transparency and could provide a deterrent to Russian technical cooperation with Iran. Others are reluctant to approve the agreement when questions remain unanswered about the Russian government's control over transfers to Iran's nuclear and missile programs.

End Notes

[1] http://www.whitehouse.gov/news/releases/2008/05/20080513-1.html.

[2] Message from the President regarding a Peaceful Nuclear Agreement with Russia, May 10, 2010, http://www.whitehouse.gov/the-press-office/message-president

[3] Under section 123.a., codified at 42 U.S.C. 2153(a)), Atomic Energy Act of 1946, ch. 724, 60 Stat. 755 (1946), as amended.

[4] Days on which either House is in a recess of more than three days (pursuant to a concurrent resolution authorizing the recess) do not count toward the total. If Congress adjourns its session *sine die*, continuity is broken, and the count starts anew when it reconvenes.

[5] See CRS Report RL34541, *Nuclear Cooperation Agreement with Russia: Statutory Procedures for Congressional Consideration and Their Implementation*, by Richard S. Beth.

[6] Nuclear cooperation includes the distribution of special nuclear material, source material, and byproduct material, to licensing for commercial, medical, and industrial purposes. These terms, "special nuclear material," "source material," and "byproduct material," as well as other terms used in the statute, are defined in 42 U.S.C. §20 14. See also CRS Report RS22937, *Nuclear Cooperation with Other Countries: A Primer*, by Paul K. Kerr and Mary Beth Nikitin.

[7] Significant nuclear cooperation includes the physical transfer of reactors, reactor components, or special nuclear material, source material, and byproduct material, under license for commercial, medical, and industrial purposes.

[8] The Atomic Energy Act also sets out procedures for licensing exports to states with whom the United States has nuclear cooperation agreements. (Sections 126, 127, and 128 codified as amended at 42 U.S.C. 2155, 2156, 2157.) Even with a 123 agreement in place, each export of nuclear material, equipment, or technology requires a specific export license or other authorization.

[9] See P.L. 99-183 and CRS Report RL33192, *U.S.-China Nuclear Cooperation Agreement*, coordinated by Shirley A. Kan.

[10] CRS Report 89-671, *U.S.-Soviet Nuclear Energy Cooperation: Time for a Full Scope Agreement?* by Warren H. Donnelly (out of print; available from the author upon request).

[11] Hearing of the Terrorism, Nonproliferation and Trade Subcommittee of the House Foreign Affairs Committee, "The Future of U.S. International Nuclear Cooperation," May 6, 2010.

[12] CRS Report R41219, The New START Treaty: Central Limits and Key Provisions, by Amy F. Woolf

[13] The United States-Russian Federation Nuclear Cooperation Agreement Act of 2008.

[14] For example, House Foreign Affairs Committee Republicans press release, August 19, 2008, http://foreignaffairs.republicans.house.gov/list/press/foreignaffairs_rep/081908123.shtml, and Joseph Biden, "Op-Ed: Russia Must Stand Down," *The Financial Times*, August 12, 2008.

[15] Statement on U.S.-Russia 123 Agreement, September 8, 2008, at http://www.state 109256.htm.

[16] http://www.whitehouse.gov/news/releases/2008/05/20080506-4.html.

[17] Message to the Congress of the United States, September 8, 2008, at http://georgewbush-whitehouse.archives.gov/ news/releases/2008/09/20080908-3.html.

[18] http://energycommerce.house.gov/Press_110/110-ltr.052208.GAO

[19] "U.S.-Russia Nuclear Agreement: Interagency Process Used to Develop the Classified Nuclear Proliferation Assessment Needs To Be Strengthened," http://www.gao.gov/new.items/d09743r.pdf

[20] "Chairman Berman responds to Administration's U.S.-Russia nuclear proposal," House Committee on Foreign Affairs Press Release, May 11, 2010.

[21] See CRS Report RL31957, Nonproliferation and Threat Reduction Assistance: U.S. Programs in the Former Soviet Union, by Amy F. Woolf.

[22] "Joint Working Group on the Development of a Bilateral Action Plan to Enhance Global and Bilateral Nuclear Energy Cooperation," at http://www.doe.gov/news/4553.htm.

[23] Text of Declaration on Nuclear Energy and Nonproliferation Joint Actions, July 3, 2007, at http://moscow.usembassy.gov/st_07032007.html.

[24] "Joint Statement by the U.S. - Russia Bilateral Presidential Commission Coordinators on Commission Progress," December 31, 2009, http://moscow.usembassy.gov/st_123109.html.

[25] "Business Groups Urge Support of US Russia Nuclear Agreement," July 11, 2008, http://www.usaengage.org/ index.php?option=com_content&task=view&id=473&Itemid=71

[26] Anton Khlopkov, "What Will a Nuclear Agreement with the United States Bring Russia?" *Security Index*, No. 2 (82), Volume 13.

[27] "Akkuyuu plant construction to begin in 2011, says Turkish energy ministry," *Platts Nucleonics Week*, May 27, 2010.

[28] See CRS Report RL34234, *Managing the Nuclear Fuel Cycle: Policy Implications of Expanding Global Access to Nuclear Power*, coordinated by Mary Beth Nikitin.

[29] According to the Atomic Energy Act, this would be considered a subsequent arrangement, under Section 131. Analysts discuss the possibility of Russia establishing an International Spent Fuel Storage Facility (ISFSF) that could accept U.S.-origin spent fuel, for example from Taiwan or South Korea, or as part of a Russian fuel leasing and return program for future nuclear power plants abroad.

[30] Import of tailings to Russia from European countries was halted in 2007 because of public protest and environmental concerns. "Russia quits uranium tailings imports over safety concerns," *RIA Novosti*, June 22, 2007. Existing contracts will be fulfilled (two with URENCO until 2009; two with EURODIF until 2014).

[31] The United States' ratification of the CSC came into effect on May 21, 2008. The CSC has not yet entered into force. Article XX.1 of the CSC states that it "shall come into force on the ninetieth day following the date on which at least 5 States with a minimum of 400,000 units of installed nuclear capacity have deposited an instrument referred to in Article XVIII." Thirteen countries have signed the CSC, but only four have ratified it.

[32] See letter from the Contractors International Group on Nuclear Liability of December 18, 2003, annexed to the testimony of Henry Sokolski to the House Committee on Foreign Affairs, June 12, 2008, at http://foreignaffairs.house.gov/110/sok061208.pdf.

[33] See also CRS Report RL3 3407, *Russian Political, Economic, and Security Issues and U.S. Interests*, coordinated by Jim Nichol.

[34] CRS Report R41219, *The New START Treaty: Central Limits and Key Provisions*, by Amy F. Woolf.

[35] Richard Lugar and Sam Nunn, "Help Russia Help Us," *New York Times*, May 30, 2008, at http://www.nytimes.com/ 2008/05/30/opinion/30lugar.html?th&emc=th.

[36] Nuclear Proliferation Assessment Statement, 2008, 2010.

[37] "Medvedev Likely to Face Problems with Iran," *RIA Novosti*, May 13, 2008, at http://en.rian.ru/analysis/20080513/ 107253545.html.

[38] CRS Report RL34544, *Iran's Nuclear Program: Status*, by Paul K. Kerr.

[39] "Prospects for a U.S.-Russian Agreement for Peaceful Nuclear Cooperation in Congress: Robert Einhorn and Jon Wolfsthal," Remarks at the Carnegie Moscow Center, April 15, 2008, http://www.carnegie.ru/en/news/78128.htm.

[40] Unclassified Report to Congress on the Acquisition of Technology Relating to Weapons of Mass Destruction and Advanced Conventional Munitions Covering 1 January to 31 December 2009. http://www.dni.gov/reports/ 2009_721_Report.pdf.

[41] Sanctions were lifted on Rosoboronexport, Dmitry Mendeleev University of Chemical Technology, the Moscow Aviation Institute and the Tula Instrument Design Bureau. Federal Register, Vol. 75, No. 98, May 21, 2010.

[42] This include sanctions under Executive Order 12938, the Iran, North Korea, Syria Nonproliferation Act, the Iran Nonproliferation Act of 2000, missile sanctions laws, etc. Available at "Nonproliferation Sanctions," State

Department website, http://www.state There are some Russian entities still under State Department sanctions for "Transfer of Lethal Military Equipment."

[43] Peter Baker and David Sanger, "Gains for Moscow in Iran deal: U.S. lifts sanctions, allows loophole to win support for resolution," *New York Times*, May 22, 2010.

[44] http://www.access

[45] Thomas Graham, "The Friend of My Enemy," *National Interest Online*, April 1, 2008, http://www.nationalinterest.org/Article.aspx?id=17266; Einhorn, Gottemoeller, McGoldrick, Poneman, Wolfsthal, "The U.S.-Russian Civil Nuclear Agreement: A Framework for Cooperation," CSIS, May 2008, http://www.csis.org/ component/option,com_csisjubs/task,view/id,4499/type, 1/.

In: U.S. Nuclear Cooperation with Other Countries
Editor: Gabriel A. Oliveira

ISBN: 978-1-61209-910-1

Chapter 5

U.S. NUCLEAR COOPERATION WITH INDIA: ISSUES FOR CONGRESS

Paul K. Kerr

SUMMARY

India, which has not signed the Nuclear Nonproliferation Treaty (NPT) and does not have International Atomic Energy Agency safeguards on all of its nuclear material, exploded a "peaceful" nuclear device in 1974, convincing the world of the need for greater restrictions on nuclear trade. The United States created the Nuclear Suppliers Group (NSG) as a direct response to India's test, halted nuclear exports to India a few years later, and worked to convince other states to do the same. India tested nuclear weapons again in 1998. However, President Bush announced July 18, 2005, he would "work to achieve full civil nuclear energy cooperation with India" and would "also seek agreement from Congress to adjust U.S. laws and policies," in the context of a broader partnership with India.

U.S. nuclear cooperation is governed by the Atomic Energy Act (AEA). P.L. 109-401, which President Bush signed into law on December 18, 2006, provides waivers of several provisions of the AEA (Sections 123 a. (2), 128, and 129). It requires that several steps occur before nuclear cooperation can proceed. On September 10, 2008, President Bush submitted to Congress a written determination that these requirements had been met. That same day, the President submitted the text of the proposed agreement, which had not yet been signed. The President also submitted a written determination (also required by the AEA) "that the performance of the proposed agreement will promote and will not constitute an unreasonable risk to, the common defense and security." In addition, President Bush submitted several documents, including classified and unclassified versions of a Nuclear Proliferation Assessment Statement, which is required by section 123 of the AEA. The Department of State also submitted a report required by P.L. 109- 401 on various aspects of the agreement.

On September 27, 2008, the House passed H.R. 7081, which approved the agreement. The Senate Committee on Foreign Relations approved identical legislation, S. 3548, September 23. The Senate passed H.R. 7081 October 1. President Bush signed P.L. 110-369 into law October 8. Secretary of State Condoleezza Rice and India's External Affairs Minister Shri Pranab Mukherjee signed the agreement October 10, and it entered into force December 6, 2008.

U.S. companies have not yet started nuclear trade with India and may be reluctant to do so if the government does not resolve concerns regarding its policies on liability for nuclear reactor operators and suppliers. India signed the Convention on Supplementary Compensation for Nuclear Damage (CSC), which has not yet entered into force, October 27, 2010. However, many observers have argued that India's Civil Liability for Nuclear Damage Bill is not consistent with the CSC.

RECENT DEVELOPMENTS

India and the United States announced July 27, 2007, that they had reached agreement on the text of a nuclear cooperation agreement.[1] P.L. 109-401, the Henry J. Hyde United States-India Peaceful Atomic Energy Cooperation Act of 2006, which President Bush signed into law December 18, 2006, allows the President to waive such an agreement with India from several requirements of the Atomic Energy Act (AEA) of 1954, as amended.

President Bush submitted the text of the proposed agreement to Congress September 10, 2008. Additionally, the President submitted a written determination (also required by the AEA) "that the performance of the proposed agreement will promote and will not constitute an unreasonable risk to, the common defense and security." In addition, President Bush submitted several documents, including classified and unclassified versions of a Nuclear Proliferation Assessment Statement (NPAS), which is required by section 123 of the AEA. The Department of State also submitted a report, which is required by section 104 of P.L. 109-401, on various aspects of the agreement.[2]

President Bush also determined that P.L. 109-401's requirements for the President to exercise his waiver authority have been met. These requirements are (1) provision of a credible separation plan for India's nuclear facilities; (2) approval by the IAEA Board of Governors of India's new nuclear safeguards agreement; (3) substantial progress toward concluding an Additional Protocol; (4) India's active support for the conclusion of a treaty to ban fissile material production for nuclear weapons; (5) India's support for U.S. and international efforts to halt the spread of sensitive nuclear fuel cycle technologies (enrichment and reprocessing); (6) India taking necessary steps to secure nuclear and other sensitive materials and technologies through adherence to multilateral control regimes (like the NSG and the Missile Technology Control Regime [MTCR]); and (7) a consensus decision by the NSG to make an exception for India.

President Bush submitted the agreement after the IAEA Board of Governors approved India's safeguards agreement August 1, 2008. The NSG decided at the end of a September 4-6 Extraordinary Plenary meeting to exempt India from the group's export guidelines.

Procedures for congressional approval of the nuclear cooperation agreement are described in both P.L. 109-401 and the AEA. According to P.L. 109-401, the agreement cannot enter into force without a joint resolution of approval from Congress. According to Section 123 b. of the AEA, the President is to submit the text of the agreement to the Senate Foreign Relations Committee and the House Committee on Foreign Affairs. The President is then to consult with the committees "for a period of not less than thirty days of continuous session." According to Section 123 d., the two committees shall, after that time, "each hold hearings on the proposed agreement for cooperation and submit a report to their respective bodies recommending whether it should be approved or disapproved." Therefore, the minimum amount of time that must elapse before Congress can vote on a joint resolution of approval is 30 days of continuous session, in addition to the amount of time Congress would take to hold hearings.

On September 27, 2008, however, by a vote of 298-117 (1 Present), the House passed H.R. 7081, which approved the agreement and waived "the provisions for congressional consideration and approval of a proposed agreement" contained in Sections 123 b. and 123 d. of the AEA. The Senate Foreign Relations Committee approved identical legislation, S. 3548, September 23. The Senate passed H.R. 7081 by a vote of 86-13 October 1. On October 8, President Bush signed P.L. 110-369, the United States-India Nuclear Cooperation Approval and Nonproliferation Enhancement Act, into law. The President's signing statement did not indicate any differences with the legislation.

According to its text, the July 2007 agreement "shall enter into force on the date on which the Parties exchange diplomatic notes informing each other that they have completed all applicable requirements for its entry into force." Secretary of State Condoleezza Rice and India's External Affairs Minister Shri Pranab Mukherjee signed the agreement October 10. On October 20, President Bush transmitted two certifications required by P.L. 110-369 in order for the two governments to exchange diplomatic notes. Washington and New Delhi exchanged diplomatic notes and the agreement entered into force December 6, 2008.

India's Nuclear Cooperation with the United States

India has stated its intention to engage in nuclear cooperation with U.S. companies. A September 10, 2008, letter from Indian Foreign Secretary Shivshankar Menon states that

> it is the intention of the Government of India and its entities to commence discussions with U.S. nuclear energy firms and conclude agreements after entry into force of the [U.S.-India] Agreement for cooperation in the construction of nuclear power units at least at two sites approved by the Government of India, which would be capable of generating a minimum of 10,000 MW.[3]

However, Menon appeared to qualify this claim, adding that such deals would be concluded

> on the basis of mutually acceptable technical and commercial terms and conditions that enable a viable tariff regime for electricity generated. It is the expectation of the Government of India that this partnership will contribute towards providing energy to India's population in a manner that takes into account affordability, sustainability of nuclear fuel resources and credibility of nuclear waste management."

New Delhi announced October 16, 2009, the specific sites that it has designated for U.S.-supplied reactors.

P.L. 110-369 requires that, before the Nuclear Regulatory Commission can issue licenses for U.S. nuclear exports to India, the President must determine and certify to Congress that New Delhi's IAEA safeguards agreement has entered into force and that India's declaration of its nuclear facilities to the agency "is not materially inconsistent with the facilities and schedule" described in a separation plan that New Delhi has provided to Washington. India's safeguards agreement entered into force May 11, 2009, and New Delhi has filed the declaration with the IAEA. President Obama submitted the required certification to Congress February 3, 2010, determining that India has satisfied the legal requirement described above.

Nevertheless, U.S. firms may be reluctant to engage in nuclear trade with India if the government does not resolve concerns regarding its policies on liability for nuclear reactor operators and suppliers.[4] India signed the Convention on Supplementary Compensation for Nuclear Damage (CSC), which has not yet entered into force, October 27, 2010. According to Menon's September 2008 letter, "it is the intention of the Indian Government to take all steps necessary to adhere to" the CSC "prior to thc commencement of international civil nuclear cooperation under" the U.S.- India agreement. In 2008, the State Department described India's decision to "become a party" to the convention as "an important step in ensuring that U.S. nuclear firms can compete on a level playing field with other international competitors" because many other countries' nuclear firms "have other liability protections afforded to them by their governments."[5]

Many observers have argued that India's Civil Liability for Nuclear Damage Bill, which both houses of India's Parliament adopted in August 20 10,[6] is not consistent with the CSC, citing the provisions that make reactor suppliers, in addition to operators, liable for damages caused by a reactor accident.[7] U.S. officials have argued that India's law should be consistent with the convention. Assistant Secretary of State Robert Blake stated in a June 9, 2010, interview with *India Abroad* that the U.S. interest is to "ensure that the bill that ultimately is enacted is compliant" with the CSC. Although Under Secretary of State William Burns described New Delhi's signing of the CSC as a "very positive step" during an October 27 press briefing, he also indicated that India will need to take additional steps in order to resolve U.S. concerns regarding India's liability policies. Shrikumar Banerjee, chair of India's Atomic Energy Commission, argued September 16 that the law is compatible with the CSC.[8]

Washington and New Delhi are also discussing necessary monitoring arrangements for U.S. nuclear exports.

Subsequent Reprocessing Arrangement

The nuclear cooperation agreement grants New Delhi consent to reprocess nuclear material transferred pursuant to the agreement, as well as "nuclear material and by-product material used in or produced through the use of nuclear material, non-nuclear material, or equipment so transferred." However, India must first "establish a new national reprocessing facility dedicated to reprocessing safeguarded nuclear material under IAEA safeguards." In addition, the United States and India must "agree on arrangements and procedures under which such reprocessing or other alteration in form or content will take place in this new facility."

The agreement also says that the two governments must begin "consultations," to be "concluded within one year," on the relevant arrangements and procedures within six months of a request from India. New Delhi made such a request February 3, 2009, and the two governments held meetings in July and October of that year.[9] The two governments completed negotiations on the agreement March 29, 2010;[10] Indian Ambassador Meera Shankar and Under Secretary of State William Burns signed the agreement July 30, 2010. India had reportedly insisted that New Delhi and Washington conclude an agreement on a reprocessing facility in India before New Delhi would sign contracts with U.S. nuclear firms.[11]

The agreement describes the "arrangements and procedures under which such reprocessing or other alteration in form or content may take place in India at two new national reprocessing facilities," which will reprocess safeguarded spent nuclear fuel from both U.S.-supplied and other reactors. The agreement also describes the procedures for U.S. officials to inspect and receive information about physical protection measures at the new facilities.

Secretary of Energy Steven Chu submitted the subsequent arrangement to Congress May 11, along with a Nuclear Proliferation Assessment Statement, a report required by the AEA, and a report and certification required by P.L. 110-369. These materials conclude that the arrangement will not increase the risk of proliferation or aid India's nuclear program. Chu also certified that the administration is pursuing "efforts to ensure" that other countries entering into similar arrangements with New Delhi do so under "similar arrangements and procedures" as the United States (see "Procedures for Subsequent Arrangements"). The arrangement would not have taken effect if Congress had adopted a joint resolution of disapproval within 30 days of continuous session; Congress did not adopt such a resolution (see "Procedures for Subsequent Arrangements"). The July 30 agreement indicates that India may construct additional facilities to reprocess fuel from U.S.-supplied reactors. However, any such facilities would require a new subsequent arrangement, which would also be submitted to Congress.

India's Nuclear Cooperation with Other Countries[12]

Since a September 2008 NSG decision to exempt India from some of its export requirements, New Delhi has negotiated nuclear cooperation agreements with NSG countries other than the United States. On September 30, 2008, India and France signed a civil nuclear cooperation agreement that includes the possible provision of nuclear reactors and nuclear fuel. It does not, however, include the transfer of enrichment or reprocessing technology, according to French Ambassador to India Jerome Bonnafont.[13] France would like India to reprocess spent nuclear fuel in an IAEA-safeguarded facility, Bonnafont said in late January 2009, but added that France would consider reprocessing the spent fuel for India.[14] According to a December 17, 2008, agreement between the French company AREVA and India's Department of Atomic Energy, AREVA agreed to supply the Nuclear Power Corporation of India with 300 metric tons of uranium. Those two companies also signed a memorandum of understanding February 4, 2009, expressing their "willingness to build up to six" nuclear reactors. Both parties "intend to discuss the elements of a commercial contract to supply" two reactors "as a first step," according to the memorandum.

Russia and India signed a nuclear cooperation agreement December 5, 2008. According to a joint declaration issued that day, "the two countries have agreed to collaborate on constructing additional nuclear power plants" and "to expand and pursue further areas for bilateral cooperation in the field of peaceful uses of nuclear energy." Notably, Russian

President Dmitry Medvedev reportedly amended in February a 1992 presidential decree on nuclear export controls in order to permit Russian nuclear exports to a country without comprehensive IAEA safeguards. However, the decree now states that

> nuclear materials, as well as technologies, equipment and special non-nuclear materials intended for their processing, utilization or production may be exported from the Russian Federation to India only if they are used in nuclear installations placed under IAEA guarantees.[15]

The Russian TVEL Corporation also reportedly signed a nuclear fuel supply contract in February 2009 with India's Department of Atomic Energy.[16]

The two countries initialed another agreement December 7, 2009, which expands on the 2008 agreement. According to a statement from India's Ministry of External Affairs, the agreement includes cooperation on research and development, the construction of additional nuclear power plants, and fuel-supply arrangements. The agreement also grants "up-front consent" for India's to reprocess spent reactor fuel and says that Russia would continue to supply fuel even if the agreement is terminated in the future.

New Delhi has also concluded other fuel-supply agreements. The Nuclear Power Corporation of India and KazAtomProm, a Kazakh national company, signed a memorandum of understanding January 24, 2009, that reportedly includes a provision for Kazakhstan to supply uranium to India under terms that remain to be determined.[17] Additionally, India and Namibia signed an "Agreement on Cooperation in Peaceful Uses of Nuclear Energy" in late August 2009. The scope and terms of the agreement are unclear, but the statement "resolved to encourage Indian investments" in the Namibian uranium sector. Similarly, India and Mongolia signed a "Memorandum of Understanding on Development of Cooperation in the Field of Peaceful Use of Radioactive Minerals and Nuclear Energy" in mid-September 2009. The agreement would reportedly enable India to explore for uranium in Mongolia.[18]

More recently, India and Argentina, according to an October 14, 2009, joint statement, agreed to "encourage and support scientific, technical and commercial cooperation for mutual benefit" in the "peaceful uses of nuclear energy." The two countries signed a nuclear cooperation agreement September 23, 2010. Furthermore, India and the United Kingdom signed a "Civil Nuclear Cooperation Declaration" February 11, 2010, which will allow for the transfer of nuclear-related technology and equipment to India. A British official explained February 13 that the agreement "opens the door to a discussion that can begin between some of the best UK-based engineering companies and the nuclear authorities in India."[19] Additionally, India and Canada signed an "Agreement between India and Canada for Co-operation in Peaceful Uses of Nuclear Energy" June 27, 2010, which provides for cooperation in nuclear reactor design and construction, as well as the "supply of uranium."[20]

BACKGROUND

The United States actively promoted nuclear energy cooperation with India from the mid-1950s, building nuclear power reactors (Tarapur), providing heavy water for the CIRUS research reactor, and allowing Indian scientists to study at U.S. nuclear laboratories. Although India was active in negotiations of the 1968 Nuclear Nonproliferation Treaty (NPT), India refused to join the treaty, arguing that it was discriminatory. India conducted a "peaceful" nuclear test in 1974, which reinforced the notion that nuclear technology transferred for peaceful purposes could be used to produce nuclear weapons.[21] Congress responded to that test by passing the Nuclear Non- Proliferation Act of 1978 (NNPA, P.L. 95-242), which imposed tough new requirements for U.S. nuclear exports to non-nuclear-weapon states—full-scope safeguards and termination of exports if such a state detonates a nuclear explosive device or engages in activities related to acquiring or manufacturing nuclear weapons, among other things.[22] The United States created the NSG, a voluntary nuclear export regime, in 1975. The NSG published guidelines in 1978 "to apply to nuclear transfers for peaceful purposes to help ensure that such transfers would not be diverted to unsafeguarded nuclear fuel cycle or nuclear explosive activities."[23]

The condition that non-nuclear-weapon states wishing to obtain U.S. nuclear exports must have full-scope safeguards created a problem for U.S. fuel supplies to India's reactors at Tarapur, which were built by U.S. firms and fueled by U.S. low-enriched uranium, pursuant to a 1963 nuclear cooperation agreement. After passage of the NNPA, the Carter Administration exported two more uranium shipments under executive order after the Nuclear Regulatory Commission (NRC) refused to approve an export license on nonproliferation conditions. Although the House voted to disapprove the President's determination, the Senate voted 46-48 on a resolution of disapproval. After 1980, all nuclear exports from the United States to India were cut off under the terms of the NNPA. France supplied fuel under the terms of the U.S. agreement with India until France also adopted a full-scope safeguards requirement in 1995. After the NSG adopted the full- scope safeguards condition in 1992, China supplied the reactor. Russia supplied fuel from 2001 to 2004.[24]

Although the NPT requires safeguards on items going to non-nuclear-weapon states, it does not explicitly prohibit nuclear commerce with states outside the NPT. In 1995, at the NPT Extension Conference, states supported the principle that non-NPT parties should not be eligible for the same kinds of assistance as NPT parties in good standing. India tested several nuclear weapons in 1998.

Global Partnership[25]

The Bush Administration had been considering a strategic partnership with India as early as 2001. Indian officials identified their growing energy needs as an area for cooperation, particularly in nuclear energy. The U.S.-India 2004 Next Steps in Strategic Partnership (NSSP) initiative included expanded cooperation in civil nuclear technology as one of three goals. Phase I of the NSSP, completed in September 2004, required addressing proliferation concerns and ensuring compliance with U.S. export controls.[26]

On July 18, 2005, President Bush announced the creation of a global partnership with India in a joint statement with Prime Minister Manmohan Singh.[27] Noting the "significance of civilian nuclear energy for meeting growing global energy demands in a cleaner and more efficient manner," President Bush said he would "work to achieve full civil nuclear energy cooperation with India" and would "also seek agreement from Congress to adjust U.S. laws and policies."

The joint statement noted that the United States

> will work with friends and allies to adjust international regimes to enable full civil nuclear energy cooperation and trade with India, including but not limited to expeditious consideration of fuel supplies for safeguarded nuclear reactors at Tarapur.

The United States committed to encouraging its partners to consider this request—a reversal in the U.S. position, which has been to ban fuel to Tarapur—and to consulting with its partners on Indian participation in ITER (collaboration on fusion research) and in the Generation IV International Forum for future reactor design.

Prime Minister Singh conveyed that India "would take on the same responsibilities and practices and acquire the same benefits and advantages as other leading countries with advanced nuclear technology, such as the United States."[28] India agreed to

- identify and separate its civilian and military nuclear facilities and programs;
- declare its civilian facilities to the IAEA;
- voluntarily place civilian facilities under IAEA safeguards;
- sign an Additional Protocol for civilian facilities;
- continue its unilateral nuclear test moratorium;
- work with the United States to conclude a Fissile Material Cut Off Treaty (FMCT);[29]
- refrain from transferring enrichment and reprocessing technologies to states that do not have them, as well as support international efforts to limit their spread;
- secure its nuclear materials and technology through comprehensive export control legislation and through harmonization and adherence to MTCR and NSG guidelines.

ISSUES FOR CONSIDERATION

The AEA requires congressional approval and oversight of peaceful nuclear cooperation agreements (details described below). As Senator Lugar has noted, "Ultimately the entire Congress ... must determine what effect the Joint Statement will have on U.S. efforts to halt the proliferation of weapons of mass destruction."[30] Congress held eight hearings in 2005 and 2006 on the global partnership and has consulted with the Administration on various aspects of the U.S.-India nuclear agreement.[31] The discussion of potential issues for consideration is drawn in part from the hearings and from the emerging debate.

Strategy vs. Tactics

The Bush Administration has described its "desire to transform relations with India" as "founded upon a strategic vision that transcends even today's most pressing security concerns."[32] There is clearly broad support for cultivating a close relationship with India, yet some Members of Congress have suggested that civil nuclear cooperation may not be the most appropriate vehicle for advancing our relationship. In a House International Relations Committee hearing on September 8, 2005, Congressman Jim Leach stated,

> I don't know any member of Congress that doesn't want to have a warming of relations with the government of India.... I also don't know many members of Congress who are pushing for the precise commitment that the administration has made.[33]

Congressman Leach suggested instead that U.S. support for a permanent seat for India on the United Nations Security Council might have been a more appropriate gesture.

Other observers outside of Congress have questioned whether U.S. energy assistance should focus on expanding nuclear power, in contrast to other energy alternatives. Henry Sokolski of the Nonproliferation Policy Education Center has argued that Indian energy needs might be better met through free market allocation, including improved efficiency. He asserts that nuclear power is the least leveraged of India's options to meet India's energy needs, given that it currently provides only 2.7% of installed electrical capacity.[34] India's projections of its nuclear energy needs are predicated on an estimated annual growth rate of 8%, which some observers believe may be unrealistic.[35] One well-known Indian commentator, Brahma Chellaney, argued in the *International Herald Tribune* that the premise that India should meet its rapidly expanding energy needs through importing nuclear power reactors was flawed. Chellaney argued that a better approach for India would be to secure clean-coal and renewable energy technologies.[36]

The Senate Foreign Relations Committee's November 2, 2005, hearing sought, among other things, to answer the question of why civil nuclear cooperation was so important to the U.S.- Indian strategic relationship. Under Secretary of State Nicholas Burns told committee members that "India had made this the central issue in the new partnership developing between our countries."[37]

Impact on U.S. Nonproliferation Policies

The Administration has characterized civil nuclear cooperation with India as a "win" for nonproliferation because it would bring India into the "nonproliferation mainstream." In short, the Administration is proposing that India should be courted as an *ally* in U.S. nonproliferation policy, rather than continue as a *target* of U.S. nonproliferation policy. According to this reasoning, India should become an ally for three reasons: past policies have not worked; India has a relatively good nonproliferation record; and India could be a useful ally in the nonproliferation regime.

Some observers, however, are concerned that India may not support U.S. nonproliferation policies sufficiently to warrant nuclear cooperation, particularly where the United States faces its greatest nuclear proliferation threat: Iran. For example, at the September 8 House International Relations Committee hearing, several Members of Congress questioned whether the United States had obtained assurances from India of its support on Iran before it issued the July 18 joint statement.

Iran

Two factors may present challenges to Indian support for U.S. policies toward Iran. First, India has a growing strategic relationship with Iran, not limited to its interest in a proposed $7.4 billion, 2800-km-long gas pipeline between Iran, Pakistan, and India. Second, India has a strong tradition of foreign policy independence, as a long-time leader of the Non-Aligned

Movement (NAM) states and as a vigorous opponent of the discriminatory nature of the Nuclear Nonproliferation Treaty.[38] One witness before the House International Relations Committee hearing on November 16, 2005, suggested that opposition from the United States on the gas pipeline project is considered to be "interference with India's autonomy in foreign relations, as well as disregard for its security and energy needs."[39]

On Iran's nuclear program, Indian officials have stated they do not support a nuclear weapons option for Tehran. However, they did not agree with the United States on the urgency of reporting Iran's nuclear program to the U.N. Security Council, which the United States has proposed since 2003, nor on the need to limit Iran's nuclear fuel cycle development. When the IAEA Board of Governors passed a resolution (GOV/2005/77) on September 24, 2005, finding Iran in noncompliance with its safeguards agreement, India voted with the United States, provoking significant domestic dissent. According to Indian Foreign Secretary Shyam Saran, India voted for the resolution and against the majority of NAM states which abstained, because it felt obligated after having pressured the EU-3 to omit reference to immediate referral to the U.N. Security Council.[40] Moreover, India explained its vote this way:

> In our Explanation of Vote, we have clearly expressed our opposition to Iran being declared as noncompliant with its safeguards agreements. Nor do we agree that the current situation could constitute a threat to international peace and security. Nevertheless, the resolution does not refer the matter to the Security Council and has agreed that outstanding issues be dealt with under the aegis of the IAEA itself. This is in line with our position and therefore, we have extended our support.[41]

On February 4, 2006, following Iran's resumption of some uranium enrichment research and development, the IAEA Board of Governors met in an emergency session and voted to report Iran's noncompliance to the U.N. Security Council.[42] India voted with the United States to report Iran, although this followed a controversial remark to the press the previous week by Ambassador David Mulford that India would have to support the United States on Iran in Vienna or the U.S. Congress would not support the peaceful nuclear cooperation agreement.[43] Several years later, New Delhi voted in support of a November 27, 2009, IAEA Board resolution that criticized Iran's failure to comply with its IAEA and U.N. Security Council obligations.

Iran may also test India's support for curtailing peaceful nuclear programs. India has always been an advocate of states' rights to develop the peaceful uses of nuclear energy and for 30 years has derided the NPT and nonproliferation policies as discriminatory. The official Iranian press agency reported Prime Minister Singh as telling President Ahmadinejad on September 22, 2005, that solutions to Iran's nuclear problem should be based on the principle that Iran as an NPT member should retain its lawful rights.[44] On September 26, 2005, Foreign Secretary Saran told the press that "With respect to Iran's right to peaceful uses of nuclear energy, that is something which we have ourselves no reservations about."[45] In September 2006, India joined in the 118-nation Nonaligned Movement (NAM) summit statement that expressed support for Iran's "choices and decisions in the field of peaceful uses of nuclear technology and its fuel cycle policies."[46]

Reported Indian Transfers of WMD-related items to Iran[47]

Concerns about India's relationship with Iran extend, for some, to the transfer of weapons of mass destruction (WMD)-related items. Entities in India and Iran appear to have engaged in very limited nuclear-, chemical-, and missile-related transfers over the years. There are no publicly available indications of activities related to biological weapons. In the early 1990s, when Iran actively sought nuclear-related assistance and technology from many foreign sources, India appears to have played only a minor role in contrast to other states. India signed an agreement in November 1991 to provide a 10 megawatt research reactor to Tehran, but cancelled under pressure from the United States. Nonetheless, India reportedly trained Iranian nuclear scientists in the 1990s.[48] More recently, India's Foreign Minister Jaswant Singh stated in December 2003 that India "has and would continue to help Iran in its controversial bid to generate nuclear energy."[49]

In September 2004, the United States imposed sanctions on two Indian nuclear scientists, Dr. Y.S.R. Prasad and Dr. C. Surendar, under the Iran Nonproliferation Act. Indian officials protested, stating that cooperation had taken place under the auspices of the IAEA Technical Cooperation program. Other reports suggest that the scientists, who had served as chairman and managing director of the Nuclear Power Corporation of India, Ltd. (NPCIL), which runs India's power reactors, passed information to Iran on tritium extraction from heavy water reactors.[50] Sanctions were lifted on Dr. Surendar in 2005.

In the chemical area, there is one confirmed transfer of 60 tons of thionyl chloride, a chemical that can be used in the production of mustard gas, from India to Iran in March 1989.[51] Other shipments in that time frame reportedly were halted under U.S. pressure. India does not appear in the CIA's unclassified nonproliferation report to Congress as a supplier of chemical-weaponsrelated exports to Iran since the report began publication in 1997. India signed the Chemical Weapons Convention in 1993 and deposited its instrument of ratification until 1996. However, in December 2005, the United States imposed sanctions on Sabero Organic Chemicals Gujarat Ltd and Sandhya Organic Chemicals Pvt. Ltd, pursuant to the Iran Nonproliferation Act of 2000. In July 2006, the United States imposed sanctions on Balaji Amines and Prachi Poly Products, chemical manufacturers, pursuant to the Iran and Syrian Nonproliferation Act.[52]

Restricting Enrichment and Reprocessing

One of India's commitments in the July 18 statement was to refrain from transferring enrichment and reprocessing technologies to states that do not already have those technologies and to support international efforts to limit their spread. As discussed in greater detail below, the NSG is considering adopting criteria for exporting enrichment and reprocessing technology.

India's External Affairs Minister Shri Pranab Mukherjee stated September 5, 2008, that:

> India will not be the source of proliferation of sensitive technologies, including enrichment and reprocessing transfers. We stand for the strengthening of the non-proliferation regime. We support international efforts to limit the spread of ENR equipment or technologies to states that do not have them.[53]

David Albright, president of the Institute for Science and International Security (ISIS), published a report on March 10, 2006, that asserted that India had potentially exported centrifuge enrichment-related technology by virtue of tendering public offers and providing blueprints for technology to interested parties.[54] ISIS followed up with a September 18, 2008, report that described Indian sales of documents related to centrifuges, as well as illicit Indian procurement of a chemical used in reprocessing.[55] It is not clear whether Indian procurement practices facilitate transfer of technology, but the U.S. nuclear cooperation agreement will have no impact on those procurement practices. Although the State Department asserted in responses to questions for the record from Senator Lugar that the United States will not engage in reprocessing or enrichment technology cooperation with India,[56] other NSG members may transfer such technology.

The NPAS notes that India, by concluding an Additional Protocol to its IAEA safeguards agreement, "will commit to reporting to the IAEA on exports of all NSG Trigger List items."

Other Priorities

In his February 11, 2004, speech, President Bush outlined several counterproliferation priorities, including expanding the Proliferation Security Initiative; strengthening laws and international controls against WMD and missile proliferation (ultimately resulting in adoption of UNSCR 1540); expanding the G8 Global Partnership; and strengthening IAEA safeguards through universal adoption of the Additional Protocol. Ambassador Joseph has noted that India's adherence to NSG and MTCR guidelines would help ensure that WMD and missile-related technologies would not be transferred. Although India's adoption of the Additional Protocol would contribute to its universalization, there are few proliferation benefits to be realized from the adoption of such a protocol in a *de facto* nuclear weapons state. Finally, although the United States reportedly has asked India to endorse PSI, that endorsement has not been forthcoming.

Impact on the Nonproliferation Regime

India's status outside the nonproliferation regime raises possible concerns that the nuclear agreement could negatively affect nuclear nonproliferation efforts. Some considerations include cohesion within the NSG, effect on non-nuclear-weapon member states of the NPT, potential missed opportunities to strengthen the nuclear nonproliferation regime, and whether U.S. nuclear cooperation might in any way assist, encourage, or induce India to manufacture nuclear weapons, in possible violation of our Article I obligation under the NPT.

NSG Cohesion

Cohesion within the NSG is critical to effective implementation of export controls. As noted earlier, the NSG has followed the U.S. lead on requiring full-scope safeguards as a condition of nuclear supply. During the September 8, 2005, hearing, House International Relations Committee Chairman Henry Hyde noted that "many of us are strong supporters of the NSG and would not want to see it weakened or destroyed." Chairman Hyde asked whether the Administration could assure the committee that

> no matter what else happens, that the administration will continue to abide by NSG guidelines, and if you are unable to gain consensus within the NSG for the amendments you need, you will not implement the new India policy in violation of NSG guidelines.

Ambassador Joseph told the committee that "we intend to take no action that would undercut the effectiveness of the NSG," and further, that the Administration did not intend to change the consensus procedure or even change the NSG full-scope safeguards condition of nuclear supply.[57] P.L. 109-401 states that the NSG should decide "by consensus to permit supply to India of nuclear items covered by" the NSG guidelines.

Dissent within the NSG could be counterproductive to achieving other objectives the United States is pursuing in nuclear nonproliferation, such as restricting the fuel cycle, disarming North Korea, and restraining Iran, all of which rely on the considerable support of friends and allies. Moreover, harmonizing export controls has played a key role in Bush counter- and nonproliferation policies in the last few years and is particularly important for interdiction efforts. U.S.-India cooperation could prompt other suppliers, like China, to justify supplying other nonnuclear-weapon states, like Pakistan. China, which joined the NSG in 2004, has reportedly favored an NSG decision based on criteria, not just an exception for India.[58] Indeed, Pakistani Foreign Secretary Shah Mahmood Qureshi said that Islamabad is interested in similar nuclear cooperation with nuclear supplier states, the *Press Trust of India* reported July 24, 2008. A July 18, 2008, letter from Pakistan's permanent representative to the IAEA similarly reflected such an interest.[59]

Russia, which only halted fuel supplies to the Indian Tarapur reactors in December 2004 at the insistence of the NSG, resumed fuel supplies to Tarapur under the guise of the safety exception, reportedly to the dismay of many NSG members.[60]

Effect on NPT Member States

India has complained for years that it has been excluded from regular nuclear commerce because of its status outside the NPT. Some observers believe this is a good thing and shows that the policy works. Others believe that a new paradigm is needed for India because it will not join the NPT as a non-nuclear-weapon state. One observer argued in a 2005 law review journal that India could join the NPT as a non-nuclear-weapon state and not give up its nuclear weapons, primarily because the NPT defines "nuclear weapon states" but does not define non-nuclear-weapon states and because the treaty does not expressly prohibit non-nuclear-weapon states from possessing nuclear weapons, just from acquiring, manufacturing, receiving transfers of or control of nuclear weapons and not to seek or receive any assistance in manufacturing nuclear weapons.[61] However, that approach would require India to stop producing fissile material for nuclear weapons and place all nuclear material (except that which is in its nuclear weapons) under IAEA safeguards.

The NPT is basically a two-way bargain. Non-nuclear-weapon states under the NPT give up the option of developing nuclear weapons in exchange for the promise of peaceful nuclear cooperation. Nuclear weapon states under the treaty commit to eventual disarmament. India, as a state outside the NPT, is bound by neither of these commitments. Some observers may see the offer of nuclear cooperation previously reserved for states under the NPT with full-scope safeguards not only as undermining the agreements made by non-nuclear-weapon states, but also the commitments made by nuclear weapon states to eventually disarm. In this view, India's continued unilateral testing moratorium is insufficient, compared with signing the Comprehensive Test Ban Treaty (CTBT), and its support for FMCT negotiations is insufficient compared with capping its nuclear weapons fissile material production now, as four of the five nuclear weapon states formally have done. Some have suggested that the absence of an Indian cap on fissile material production for weapons may make it difficult for China to declare it has halted fissile material production for weapons. Others have suggested that, absent a cap on fissile material production, it would be difficult to ensure that peaceful nuclear cooperation was not indirectly assisting or encouraging India's nuclear weapons program.

The proliferation shocks of the 1990s, when the Iraqi and North Korean clandestine nuclear weapons programs surfaced, led to the strengthening of the NPT and export control regimes. At the 1995 NPT Review and Extension Conference, NPT parties affirmed the NSG's decision to require full-scope safeguards for nuclear exports, supporting the principle that non-NPT parties should not be eligible for the same kinds of assistance as NPT parties in good standing. At the 2000 conference, NPT parties again supported that principle. According to the U.S. ambassador to the conference at that time, "reinforcement of this guideline is important given some who have questioned whether this principle should be relaxed for India and Pakistan, which have not accepted full-scope IAEA safeguards. The answer from NPT parties is clearly no."[62]

In the past 10 years, virtually all states agreed to strengthen the nonproliferation regime, sacrificing some sovereignty by allowing additional, intrusive inspections under the Additional Protocol. In the wake of revelations in 2004 about Pakistani scientist A.Q. Khan's nuclear black market sales, non-nuclear-weapon states under the NPT are also being asked to consider further restrictions on their sovereignty by voluntarily restricting their access to sensitive nuclear technologies like uranium enrichment and reprocessing. If some states view the U.S.-Indian nuclear cooperation agreement as a breach of faith in the basic bargain of the NPT, they might be less inclined to accept additional sacrifices, to the detriment of the nonproliferation regime.

Missed Opportunities

Ambassador Joseph described the nuclear initiative as representing "a substantial net gain for nonproliferation. It is a win for our strategic relationship, a win for energy security, and a win for nonproliferation." Ambassador Joseph said he was "convinced that the nonproliferation regime will emerge stronger as a result."[63]

However, some observers have suggested the United States asked for too little. For example, Fred McGoldrick, Harold Bengelsdorf, and Lawrence Scheinman argued in the October 2005 issue of *Arms Control Today* that

> It is open to serious doubt whether the proposed Indian concessions were significant enough to justify the accommodations promised by the United States and whether the steps the United States and India agreed to take in the civil nuclear area will, on balance, be supportive of global nonproliferation efforts.... If the Bush Administration is able to implement the joint declaration without significant modification, it will have given the Indians a great deal— acknowledgment as a de facto nuclear weapon state and access to the international nuclear energy market—in return for largely symbolic concessions in the nonproliferation area.[64]

Robert Einhorn of the Center for Strategic and International Studies told Members of the House International Relations Committee on October 26, 2005, that several of the steps pledged by India are "simply reaffirmations of existing positions."[65] The Indian embassy itself has downplayed the depth and breadth of its nonproliferation commitments, describing all but its safeguards commitments under the July 2005 statement in the following way:

> A number of existing policies were also reiterated by India, among them a unilateral moratorium on nuclear testing, working towards conclusion of a multilateral Fissile Material Cut-off Treaty, non-transfer of enrichment and reprocessing technologies, securing nuclear materials and technology through export control, and harmonisation with MTCR and NSG guidelines.[66]

India has had a self-imposed nuclear test moratorium for years, although supporters of this agreement note that this agreement would bind India bilaterally to honoring that pledge. If the NSG used a similar criterion in approving exports, it could further strengthen that pledge. India has supported FMCT negotiations for years, despite continuing to produce fissile material for use in nuclear weapons. Since the pace of FMCT negotiations has been glacial, support for negotiations could allow India to continue producing fissile material indefinitely. Moreover, the draft treaty on FMCT tabled by the United States in Geneva on May 18, 2006, would not require India's accession for the treaty to enter into force, thus lessening the pressure on India to join. Prime Minister Singh told his Parliament on August 17, 2006, that "India is willing to join only a non discriminatory, multilaterally negotiated and internationally verifiable FMCT."[67] Unfortunately, this conflicts with the U.S. draft proposal, which contains no verification measures.

The most far-reaching of the commitments is to separate civilian and military facilities, declare civilian facilities, and place them under safeguards. Administration officials have pointed to this aspect of the agreement as a nonproliferation "plus." Yet, allowing India broad latitude in determining which of its facilities to put under international safeguards is a privilege accorded currently only to nuclear weapon states under the NPT. Although the United States "in no way recognizes India as an NPT nuclear weapons state," excluding military facilities from inspections is a tacit recognition of their legitimacy.

IAEA Director General Dr. Mohamed ElBaradei said that he has "always advocated concrete and practical steps towards the universal application of IAEA safeguards."[68] In remarks to the Carnegie Endowment's Nonproliferation Conference in November 2005, Dr. ElBaradei cited additional safety benefits of putting more Indian facilities under safeguards. However, it should be noted that the NSG already has an exception to its full-scope safeguards requirement for safety-related items.

The Administration has asserted that India has an "exceptional" record of nonproliferation and, despite a few isolated sanctions, most of the evidence supports the view that India has exercised restraint in export controls.[69] As such, however, New Delhi's promise to refrain from transferring enrichment and reprocessing technologies to states that do not have them, as well as its promise to adhere to NSG guidelines, may be little more than a formality.

Many observers have noted that there are no measures in this global partnership to restrain India's nuclear weapons program. Many have suggested that the United States should have asked New Delhi to halt fissile material production for weapons. Ambassador Joseph stated that the United States remains "committed to achieving Indian curtailment of fissile material production, and we have strongly encouraged a move in this direction. We stand willing to explore options that might serve this objective, but we will not insist on it for purposes of this civil nuclear initiative."[70] Indian officials, on the other hand, have taken pains to point out that "there is no commitment at all to cease production of fissile material ahead of the conclusion of such a multilateral [FMCT] treaty."[71] Prime Minister Singh told the Parliament in August 2006 that "our position on this matter is unambiguous. We are not willing to accept a moratorium on the production of fissile material."[72] Other observers have noted that although India committed to a test ban, it did not commit to signing the CTBT. Still other observers have suggested that if India insists on being treated as a nuclear weapon state, it should undertake responsibilities similar to those of the other nuclear weapon states, for example, placing fissile material excess to defense needs under safeguards. Many believe that real limits on India's nuclear weapons program would constitute a "win" for nonproliferation.

U.S. NPT Article I Obligations/Aiding India's Nuclear Arsenal

Article I of the NPT states that

> each nuclear-weapon State Party to the Treaty undertakes not to transfer to any recipient whatsoever nuclear weapons or other nuclear explosive devices or control over such weapons or explosive devices directly, or indirectly; and not in any way to assist, encourage, or induce any non-nuclear-weapon State to manufacture or otherwise acquire nuclear weapons or other nuclear explosive devices, or control over such weapons or explosive devices.

Given that India will continue to make nuclear weapons, but is considered under the NPT to be a non-nuclear-weapon state, the question arises as to whether U.S. assistance might in any way violate Article I.[73] In testimony before the House International Relations Committee, David Albright of ISIS stated that "without India halting production of fissile material for its nuclear weapons programs, nuclear assistance, particularly any in the areas involving the fuel cycle, would likely spill over to India's nuclear weapons program."[74]

Three areas raise potential concerns: whether the separation plan is adequate to ensure that cooperation does not in any way assist in the development or production of nuclear weapons; whether cooperation confers nuclear weapons state status on India, with an unintended consequence of encouraging the Indian nuclear weapons program; and whether opening up the international uranium market frees up India's domestic uranium for use in its weapons program.

Separation Plan Credibility

Administration officials have defended the separation plan as credible and defensible because it covers more than just a token number of Indian facilities, provides for safeguards in perpetuity, and includes upstream and downstream facilities.[75] However, although IAEA safeguards ensure that nuclear material is not diverted, there are no procedures or measures in place to ensure that information, technology and know-how are not transferred from the civil sector to the military sector. This issue, which the September 2008 NPAS does not appear to address, could become a key loophole. For example, U.S. assistance to one of the eight indigenous power reactors, whether focused on nuclear safety, improving operational efficiency, or extending its lifetime, could easily be applied by Indian personnel to one of the similar, but unsafeguarded indigenous power reactors. Some Indian commentators have suggested that the United States has little technology to offer India, and others have doubted whether U.S. assistance would be provided to those indigenous power reactors.

Conferring Nuclear Weapon-State Status

A second area that raises concerns is whether nuclear cooperation confers nuclear weapon state status on India, which could encourage its weapons program. Senator Lugar noted in a hearing on November 2, 2005, that

> prior to the July 18 joint statement India had repeatedly sought unsuccessfully to be recognized as an official nuclear weapons state, a status the NPT reserves only for the United States, China, France, Russia and the United Kingdom. Opponents argue that granting India such status will undermine the essential bargain that is at the core of NPT, namely, that only by foregoing nuclear weapons can a country gain civilian nuclear assistance.

Dr. Ashton Carter, testifying at that SFRC hearing, stated that:

> India obtained *de-facto* recognition of its nuclear weapons status. The United States will behave, and urge others to behave, as if India were a nuclear weapons state under the NPT. We won't deny it most civil nuclear technology or commerce. We won't require it to put all of its nuclear facilities under IAEA safeguards—only those it declares to be civil. Beyond these technicalities, nuclear recognition confers an enormous political benefit on India.

Secretary Rice, in response to a question for the record in April 2006 on India's nuclear weapon state status, stated that "while India has nuclear weapons and we must deal with this fact in a realistic, pragmatic manner, we do not recognize India as a nuclear weapon state or seek to legitimize India's nuclear weapons program." However, other officials' statements appear to lend more support to India. Under Secretary of State Nicholas Burns told reporters on March 2, 2006, that "India is a nuclear weapons power, and India will preserve part of its nuclear industry to service its nuclear weapons program."[76]

IAEA Director General ElBaradei argues that the U.S.-India deal is "neutral" because "it does not confer any 'status,' legal or otherwise, on India as a possessor of nuclear weapons."[77] Nonetheless, the successful U.S. effort to create an exemption in its nuclear cooperation law for New Delhi has placed India in the company of only four other nations—the United Kingdom, France, China, and Russia—all *de jure* nuclear weapon states. Many observers believe that this legitimizes India's nuclear weapons program by providing *de facto* recognition. Indian official statements repeatedly have used the term "advanced nuclear states" as synonymous with nuclear weapon states; Prime Minister Singh told the Parliament in August 2006 that

> The July Statement did not refer to India as a Nuclear Weapons State because that has a particular connotation in the NPT, but it explicitly acknowledged the existence of India's military nuclear facilities. It also meant that India would not attract full-scope safeguards such as those applied to non-nuclear weapon states that are signatories to the NPT and there would be no curbs on continuation of India's nuclear weapon related activities. In these important respects, India would be very much on par with the five Nuclear Weapon States who are signatories to the NPT. Similarly, the Separation Plan provided for an India specific

safeguards agreement with the IAEA with assurances of uninterrupted supply of fuel to reactors together with India's right to take corrective measures in the event fuel supplies are interrupted. We have made clear to the US that India's strategic programme is totally outside the purview of the July Statement, and we oppose any legislative provisions that mandate scrutiny of our nuclear weapons programme or our unsafeguarded nuclear facilities.

Offsetting India's Uranium Shortage

Finally, critics of U.S.-Indian civil nuclear cooperation have argued that giving India access to the international uranium market would free up New Delhi's domestic uranium resources for use in its weapons program.[78] India has limited indigenous uranium reserves and apparently has difficulty producing reactor fuel from these reserves. New Delhi has reportedly reduced its power reactors' electricity output because of fuel shortages.[79] Since the NSG's 1992 adoption of the full- scope safeguards condition for exports, New Delhi has not had access to the international uranium market.[80] Foreign Secretary Saran pointed out on February 18, 2008, that "a major expansion in nuclear power in the medium term" would require India to import "higher capacity reactors and uranium fuel."[81] Similarly, the NPAS states that "India must import fuel, reactors, and other technologies ... to meet its nuclear electricity-generating targets."

India's current fuel situation means that New Delhi cannot produce sufficient fuel for both its nuclear weapons program and its projected civil nuclear program. A panel of nuclear experts concluded in 2006 that

> India's production of weapon grade plutonium is currently constrained by the competing demands of India's nuclear-power reactors for its limited domestic supply of natural uranium. If India could import fuel for its civilian nuclear reactors, it could use more domestic uranium for the production of weapon materials.[82]

Acknowledging the country's uranium limitations, a leading Indian nuclear strategist K. Subrahmanyam suggested December 12, 2005, in *The Times of India*, that New Delhi should use imported nuclear fuel to preserve its domestic uranium reserves for nuclear weapons:

> Given India's uranium ore crunch and the need to build up our minimum credible nuclear deterrent arsenal as fast as possible, it is to India's advantage to categorize as many power reactors as possible as civilian ones to be refueled by imported uranium and conserve our native uranium fuel for weapon-grade plutonium production.[83]

Proponents of U.S.-Indian nuclear cooperation have made several arguments regarding the Article I issue. None of these claims, however, refute the fact that India's access to the international uranium market will result in more indigenous Indian uranium available for weapons because it will not be consumed by New Delhi's newly safeguarded reactors.[84] For example, proponents of the deal argue that India already has sufficient uranium to increase its nuclear arsenal and that New Delhi does not plan to increase that arsenal. The NPAS states that "India is capable of maintaining and expanding its existing nuclear arsenal within the limit of its indigenous resources and capabilities." Secretary Rice made a similar argument during an April 5, 2006, House International Relations Committee hearing. But these arguments do not address the constraints India faces in pursuing its civilian and military nuclear programs using only indigenous uranium.

The NPAS also addresses New Delhi's intentions regarding its future nuclear arsenal. First, the statement says that Washington has "no evidence indicating that India plans to use additional domestic uranium reserves in its nuclear weapons programs as a consequence of implementing" the nuclear cooperation agreement. Second, the NPAS argues that New Delhi has "a posture of nuclear restraint," citing India's stated no-first-use nuclear weapon policy and need for only a "credible minimum deterrent." India, however, has never defined what it means by such a deterrent; Saran stated during an April 2006 television interview that "there is no responsibility on part of India to declare what its minimum deterrent is."[85] Moreover, New Delhi's intentions are irrelevant to determining U.S. compliance with its Article I obligations.

Furthermore, the NPAS appears to argue that the agreement may reduce India's potential to produce additional fissile material for weapons, because the additional indigenous Indian reactors placed under safeguards "will no longer be available for this purpose." However, the statement characterizes the reactors' potential to produce fissile material for weapons as "theoretical." And, as noted above, India is not obligated to place future reactors under safeguards.

The State Department report submitted with the agreement notes that IAEA safeguards will prevent India from using its civil nuclear facilities for its nuclear weapons program. It also describes the ways in which U.S. export control policies prevent exported U.S. dual-use nuclear technologies from being used for military purposes.[86]

In the view of many nonproliferation analysts, the key to ensuring that civilian nuclear cooperation does not assist India's weapons program is to insist on New Delhi halting its fissile material production for weapons. That would narrow the area of concern to technology transfer to the weapons and delivery systems themselves, rather than fissile material production in reactors, enrichment facilities, and reprocessing plants. Among others, Henry Sokolski suggested in the *Wall Street Journal* that "if we want to keep this aid from freeing up India's domestic nuclear resources to make more bombs ... we have to get serious about India capping its nuclear weapons program."

STEPS REQUIRED BY P.L. 109-401

As noted above, P.L. 109-401 allows the President to exempt a U.S. nuclear cooperation agreement with India from several AEA requirements. Before such an agreement can enter into force, however, the President must first certify that several steps have been completed. President Bush did so September 10, 2008. After the President signs H.R. 7081, which both the House and Senate have passed, Washington and New Delhi must, according to the agreement's text, "exchange diplomatic notes informing each other that they have completed all applicable requirements" in order for the agreement to enter into force.

Separation Plan and Safeguards[87]

U.S. and Indian officials agreed on India's separation plan in March 2006. The key elements of that plan are:[88]

- Eight indigenous Indian power reactors will be placed under an India-specific safeguards agreement, bringing the total number of power reactors under safeguards to 14 of 22 (six are already under safeguards);[89]
- Future power reactors may also be placed under safeguards, if India declares them as civilian;
- Some facilities in the Nuclear Fuel Complex (e.g., fuel fabrication) will be specified as civilian in 2008; and
- Nine research facilities and three heavy water plants would be declared as civilian, but are "safeguards-irrelevant."

The following facilities and activities were not on the separation list:

- Eight indigenous Indian power reactors,
- Fast Breeder test Reactor (FTBR) and Prototype Fast Breeder Reactors (PFBR) under construction,
- Enrichment facilities,
- Spent fuel reprocessing facilities (except for the existing safeguards on the Power Reactor Fuel Reprocessing (PREFRE) plant),
- Research reactors: CIRUS (which will be shut down in 2010), Dhruva, Advanced Heavy Water Reactor,
- Three heavy water plants, and
- Various military-related plants (e.g., prototype naval reactor).

The separation plan stated that India would begin placing facilities under safeguards in 2006 and complete the process in 2014. However, since the IAEA did not approve New Delhi's safeguards agreement until 2008, India will need to update that timeline. Acting Deputy Assistant Secretary of State for International Security And Nonproliferation Richard Stratford told the Senate Foreign Relations Committee September 18, that New Delhi stands by its initial plan to bring its facilities under safeguards by 2014.

India's Implementation Document noted that facilities were excluded from the civilian list if they were located in a larger hub of strategic significance, even if the facilities themselves were not normally engaged in activities of strategic significance, calling into question whether the plan really will result in a "separation" of civilian and military facilities. Moreover, the plan stated that electricity grid connectivity is not relevant to the separation exercise and that grid connectivity would be necessary "irrespective of whether the reactor concerned is civilian or not civilian." This means that "military" reactors will continue to provide civilian electricity.

Another question that arises is whether India, in the absence of full-scope safeguards, can provide sufficient confidence that U.S. peaceful nuclear technology will not be diverted to nuclear weapons purposes, as many believe it was in 1974.[90] In response to a question for the record submitted by Senator Lugar on April 5, 2006, on whether exports of nuclear material or reactors from the United States would in any way assist India's nuclear weapons program, the Administration noted that "any items sent to India would be subject to safeguards, and implementation of the Additional Protocol would provide further assurances of the non-diversion of such items or material."[91] However, the Additional Protocol provides assurances of absence of undeclared activities, rather than of the non-diversion of safeguarded items, contrary to Secretary Rice's assertions.

Administration officials repeatedly have stressed that India's separation plan must be credible, transparent, and defensible from a nonproliferation standpoint,[92] and that "the resultant safeguards must contribute to our nonproliferation goals."[93] To those observers who interpreted that statement to mean that a separation plan would need to take into account India's past commitments (e.g., use of purportedly "peaceful" nuclear reactors like CIRUS to produce plutonium for nuclear weapons) and the impact on its nuclear weapons program (e.g., capping India's fissile material production), the separation plan may not appear credible. To those observers who interpreted "credible" to mean that all power reactors that supplied electricity would be declared civilian because they have a civilian use, the separation plan also may not appear credible. Secretary Rice has stressed, however, that more reactors under safeguards means more transparency, more physical security, better nuclear safety, and therefore increased safety for the United States.[94] Even so, some observers may argue that types of facilities safeguarded are critical in assessing whether the plan is defensible from a nonproliferation standpoint. For example, in terms of preventing terrorist access to fissile material, safeguarding facilities like reprocessing and enrichment plants and breeder reactors could be viewed as providing a significant nonproliferation benefit because the materials produced by these plants are a few steps closer to potential use in a bomb. In addition, safeguards on enrichment, reprocessing plants, and breeder reactors would support the 2002 U.S. National Strategy to Combat Weapons of Mass Destruction, in which the United States pledged to "continue to discourage the worldwide accumulation of separated plutonium and to minimize the use of highly-enriched uranium."[95]

India-IAEA Discussions/Domestic Opposition[96]

After the United States and India concluded the nuclear agreement last July, New Delhi delayed beginning talks with the IAEA about a safeguards agreement because of domestic opposition from Communist and other leftist parties, known as the Left Front.[97] Until recently, the United Progressive Alliance government, led by Prime Minister Singh, depended on those parties' support in order to stay in power. In India, the executive can enter into international agreements without parliamentary approval, but the Left Front threatened to withdraw its support if the government went ahead with the safeguards discussion. Indian officials had indicated multiple times that the government would not risk prompting early elections in order to push the deal through. In November 2007, the Left Front agreed to allow the government to engage in discussions with the IAEA. The talks were announced November 21, and the two parties subsequently met five times.

New Delhi had indicated that, once a safeguards text has been agreed upon with the IAEA Secretariat, the government would seek approval from an ad hoc political committee (which includes the Communists) before proceeding further with the agreement. Speaking before a Calcutta audience February 3, 2008, External Affairs Minister Mukherjee said that when "the draft agreement [with the IAEA] is ready it will be brought back to the United Progressive Alliance (UPA)-Left Coalition committee for its approval and suggestion."[98] Similarly, Communist Party of India (Marxist) General Secretary Prakash Karat stated the previous November that "we have come to an understanding that the government can go to the IAEA secretariat. But the outcome of the talks should be brought to the committee before moving to the IAEA board of governors."[99] The committee last met June 25, 2008. The government, however, never presented the text of the safeguards agreement to the committee, and at New Delhi's request, the IAEA Secretariat circulated the draft text (GOV/2008/3 0)[100] July 9, 2008, to the agency's board.

Pakistan, along with several unidentified board members, had "voiced strong reservations" about the safeguards agreement, according to a July 24, 2008, *Nucleonics Week* article,[101] but the Board of Governors approved the agreement by consensus August 1, 2008.

Four Left Front parties withdrew their support for the coalition government July 9, 2008, shortly after Singh announced that India would approach the IAEA board. However, the coalition government narrowly won a July 22, 2008, vote of confidence, staving off the threat of early elections. Karat stated September 7, 2008, that the Left Front would only support a government that would terminate the nuclear agreement with the United States.[102] However, despite the loss of the Left Front's support, the coalition government considerably bolstered its standing in April- May 2009 parliamentary elections, thereby obtaining a more stable coalition. The opposition Bharatiya Janata Party has expressed its opposition to the deal and has stated that it would attempt to renegotiate it if the party regains power. [103]

India's Safeguards Agreement [104]

IAEA Director General ElBaradei described India's safeguards agreement as an "umbrella agreement" that allows for any facility identified by New Delhi in the future to become subject to safeguards. Since New Delhi has committed to place additional reactors under safeguards, Elbaradei added, concluding an umbrella agreement was more efficient than negotiating different agreements for each facility. India signed the agreement February 2, 2009, and it entered into force May 11, 2009.

The safeguards agreement requires India to provide the IAEA with a declaration of its nuclear facilities. India is to implement this provision in a two-step process. First, according to paragraph 13 of the agreement, New Delhi will provide a declaration of nuclear facilities that it intends to place under safeguards in the future. Second, according to paragraph 14 of the agreement, India is to notify the IAEA when specific facilities are to be safeguarded. Those facilities will be placed on an Annex to the agreement. India will also have to notify the agency of imported items that are required to be safeguarded, but these will not be listed in the Annex.

On July 25, 2008, India provided to the IAEA a document—a copy of New Delhi's 2006 separation plan—containing a list of its nuclear facilities.[105] A State Department official said that India's submission of this document did not constitute submission of the declaration required by paragraph 13 of the safeguards agreement, but did satisfy P.L. 109-401's requirement regarding New Delhi's declaration.[106] India subsequently provided the declaration pursuant to paragraph 13 of the agreement, but exactly when it did so is unclear. In October 2009, New Delhi provided the IAEA, "in accordance with" paragraph 14 of the agreement, with a list of 14 nuclear facilities that are to be placed on the safeguards agreement's Annex.[107] Six nuclear reactors listed in the separation plan were not included in this submission, but India intends, in accordance with the separation plan, to place them under safeguards by 2014.[108]

An Indian official previously indicated that New Delhi's placement of nuclear facilities under safeguards was contingent on India's conclusion of nuclear supply agreements with other countries. Anil Kakodkar, chairman of India's Atomic Energy Commission, said in a

July 20, 2008, interview that India's identification of "any facility as civilian is conditional on that facility benefitting from full civil nuclear cooperation" with other countries.[109]

Some observers have expressed concerns about the agreement's preamble, which contains language suggesting that India could withdraw nuclear facilities or fuel from safeguards if New Delhi so chooses. For example, the preamble states that India "may take corrective measures to ensure uninterrupted operation of its civilian nuclear reactors in the event of disruption of foreign fuel supplies." New Delhi has not defined "corrective measures," although Kakodkar described them in July 2008 as "unspecified sovereign actions."[110]

The preamble also states that

> [a]n essential basis of India's concurrence to accept Agency safeguards is ... support for an Indian effort to develop a strategic reserve of nuclear fuel to guard against any disruption of supply over the lifetime of India's reactors.

New Delhi may want such a stockpile to hedge against a cut-off of fuel supplies in the event that, for example, India tests a nuclear weapon.

However, ElBaradei stated August 1, 2008, that the agreement's specific termination clauses "override any general clauses in the agreement." Additionally, the State Department stated in January 2008 responses to Questions for the Record submitted by the House Committee on Foreign Affairs that New Delhi "has expressed its view that for purposes of implementing the U.S.-India Agreement," IAEA safeguards "can and should be regarded as being 'in perpetuity.'"[111]

It is also worth noting that, if India were to terminate IAEA safeguards on U.S. nuclear exports (or special nuclear material produced from or with such exports), section 123 a. (1) of the AEA requires that fall-back safeguards be maintained on those exports. The nuclear cooperation agreement states that

> safeguards will be maintained with respect to all nuclear materials and equipment transferred pursuant to this Agreement, and with respect to all special fissionable material used in or produced through the use of such nuclear materials and equipment, so long as the material or equipment remains under the jurisdiction or control of the cooperating Party.

The State Department noted that if IAEA safeguards "fail to be applied," the two countries "must enter into arrangements for alternative measures to fulfill" the above requirement.[112] Furthermore, also stated that "it would not be consistent with the proposed agreement text for ... corrective measures to detract from the applicability" of safeguards to the relevant nuclear items "including after termination or expiration of the agreement,"[113] according to the Department's January 2008 responses.

NSG Support

Following formal and informal U.S. consultations with NSG members, the United States presented during a March 2006 Consultative Group meeting a draft decision for potential discussion during the NSG plenary in May 2006.[114] That draft sought an exception for India to the NSG requirements of full-scope safeguards, notwithstanding the exceptions for safety assistance and for those agreements signed before the full-scope safeguards requirement came into effect in 1992. It did not contain any restrictions on enrichment or reprocessing cooperation, nor on heavy water or HEU or plutonium sales.[115]

The United States subsequently developed a second draft decision, which incorporated the suggestions of supporting NSG members.[116] After revising that draft following consultations with New Delhi, Washington submitted it to the NSG chair in early August 2008.[117] The second version did not contain any additional restrictions on India. Indeed, it weakened one section of the 2006 draft which stated that NSG members could engage in nuclear trade with New Delhi if "the participating Government intending to make the transfer is satisfied that India continues to fully meet all" of its nonproliferation and safeguards commitments. The new draft stated only that

> Participating Governments shall maintain contact and consult through regular channels on matters connected with the implementation of the Guidelines, taking into account relevant international commitments and bilateral agreements with India.

The NSG considered the new draft decision during an August 21-22 Extraordinary Plenary meeting and decided during a similar meeting held September 4-6, 2008, to exempt India from some of its export guidelines. This decision means that members' decisions to export previously restricted nuclear items to India are now governed by individual governments' policies.[118]

Although several countries advocated that certain conditions (such as an explicit ban on the transfer of enrichment and reprocessing technology, as well as a provision that nuclear supplies to India would end if New Delhi were to test a nuclear weapon) be included in an exemption for India, the final language contains no such explicit conditions.[119] Instead, it states that the exemption is "based on" Indian commitments and actions, which are essentially the same as the requirements in P.L. 109-401. The NSG statement also notes that India has agreed to continue its "unilateral moratorium on nuclear testing."

The NSG agreed to exempt India from the portions of its guidelines that require India to have full-scope IAEA safeguards "provided that transfers of sensitive exports [enrichment and reprocessing technology] remain subject to paragraphs 6 and 7 of the Guidelines."

The relevant portions of those paragraphs state that

> suppliers should exercise restraint in the transfer of sensitive facilities, technology and material usable for nuclear weapons or other nuclear explosive devices. If enrichment or reprocessing facilities, equipment or technology are to be transferred, suppliers should encourage recipients to accept, as an alternative to national plants, supplier involvement and/or other appropriate multinational participation in resulting facilities.... [and]

> For a transfer of an enrichment facility, or technology therefor, the recipient nation should agree that neither the transferred facility, nor any facility based on such technology, will be designed or operated for the production of greater than 20% enriched uranium without the consent of the supplier nation, of which the IAEA should be advised.

Ireland reportedly stated after the NSG decision that, "on the basis of consultations during the meeting, it 'understands that no [NSG member] currently intends to transfer to India any facilities, equipment, materials, or technology related to the enrichment of uranium, or the reprocessing of spent fuel.'"[120] Under Secretary of State for Arms Control and International Security John Rood made a similar assertion during a September 18, 2008, Senate Foreign Relations Committee hearing.

Berman explained September 26 that he chose to support H.R. 7081, the bill approving the agreement, partly because Secretary Rice "made a personal commitment" to him that "the United States will make its 'highest priority' at the November meeting of the Nuclear Suppliers Group the achievement of a decision by all of the nuclear suppliers to prohibit the export of enrichment and reprocessing equipment and technology" to non-NPT states. Asked the same day about Berman's statement, Rice told Reuters that the United States would advocate for "strict limits" on the export of such technology.

The NSG is considering adopting criteria for exporting enrichment and reprocessing technology. President Bush stated in February 2004 that NSG members should refuse to sell "enrichment and reprocessing equipment and technologies to any state that does not already possess full-scale, functioning enrichment and reprocessing plants." The Administration has since shifted its position and proposed export criteria of its own. France first introduced such criteria in 2004.[121]

According to the September 2008 NSG statement, participating governments will meet "and act in accordance with paragraph 16 of the [group's] Guidelines" if one or more members "consider that circumstances have arisen which require consultations." Paragraph 16 provides a list of potential steps for NSG members to take if

> one or more suppliers believe that there has been a violation of supplier/recipient understanding resulting from these Guidelines, particularly in the case of an explosion of a nuclear device, or illegal termination or violation of IAEA safeguards by a recipient.

Under such circumstances, NSG members could agree to cut off nuclear supplies; indeed, New Zealand stated September 6 that "in the event of a nuclear test by India, this exemption will become null and void." However, the NSG would have to agree by consensus to cut off nuclear exports.

Several NSG governments indicated in statements after the vote that a September 5, 2008, statement from Mukherjee describing India's "stand on disarmament and nonproliferation" played a decisive role in persuading them to agree to the exemption.[122] However, Mukherjee simply reiterated previous Indian policies and articulated no new commitments. For example, he stated that India "remain[s] committed to a voluntary, unilateral moratorium on nuclear testing," but New Delhi is already committed to such a moratorium. Furthermore, Congress Party spokesperson Manis Tiwari stated September 4 that "[i]f the need arises and if it is in our national interest, we are ready to test not once, not twice but a hundred times."[123]

Other Required Steps

Additional Protocol

Section 104 (b) (3) of P.L. 109-401 requires a presidential determination that "India and the IAEA are making substantial progress toward concluding an Additional Protocol consistent with IAEA principles, practices, and policies that would apply to India's civil nuclear program." According to the State Department, ElBaradei concluded September 10 that India has made substantial progress toward concluding such a protocol. The IAEA Board of Governors approved the protocol March 3, 2009, and India signed it May 15, 2009.

As previously noted, Additional Protocols, which are based on a Model Additional Protocol, are designed to augment the IAEA's ability to detect undeclared nuclear activities in a member-state. Since New Delhi has nuclear weapons and is keeping some of its nuclear facilities outside of safeguards, "there are bound to be important differences between" India's Additional Protocol and the Model Protocol, according to the NPAS. Indeed, India's Additional Protocol does not contain most of the Model Protocol's provisions, requiring only that India provide the IAEA with information about its nuclear exports.

Declaration of Nuclear Facilities

Section 104 (b) (1) requires that India file "a declaration regarding its civil facilities and materials with the IAEA." As noted above, a State Department official said that the India's submission of this plan does not constitute submission of the declaration required by the safeguards agreement, but does satisfy P.L. 109-401's requirement.[124] India has, as noted, submitted its declaration to the IAEA. New Delhi is neither required to place any new facilities under safeguards, nor to adhere to the separation plan. However, an Indian facility must be placed under safeguards in order for it to receive foreign assistance.

Harmonization with NSG/MTCR Guidelines and Adherence to NSG Guidelines

Section 104 (b) (6) (B) requires a presidential determination that India has harmonized "its export control laws, regulations, policies, and practices with the guidelines and practices" of the MTCR and the NSG. Section 104 (b) (6) (B) requires a determination that New Delhi adhere to the NSG guidelines.

The State Department report assesses that India has harmonized its export control laws with the NSG and MTCR guidelines "up through the 2005 revisions, and has the means in place to make future updates to its guidelines and control lists."[125] New Delhi stated its adherence to the NSG guidelines in a September 8, 2008, letter to the IAEA and its adherence to the MTCR guidelines in a September 9, 2008, letter to the MTCR point of contact.[126]

According to the September NSG statement, the group's chair (which rotates each year) "is requested to consult with India regarding changes to and implementation of" the group's guidelines. This provision gives New Delhi what is essentially a non-binding consultative role in formulating changes to the guidelines. India had been reluctant to adhere to the guidelines because they sometimes change and New Delhi, as a non-member, will not be able to participate in the group's decisions regarding such changes.[127]

The Atomic Energy Act and Consultations with Congress

Under existing law (Atomic Energy Act [AEA] of 1954; P.L. 95-242; 42 U.S.C. § 2153 et seq.) all significant nuclear cooperation requires an agreement for cooperation. [128] The Nuclear Non- Proliferation Act of 1978 (NNPA) amended the Atomic Energy Act of 1954 to include, among other things, a requirement for full-scope safeguards for significant nuclear exports to nonnuclear-weapon states.[129]

At issue are the requirements for full-scope nuclear safeguards contained in Section 123 a. (2) for approval of an agreement for cooperation and in Section 128 for licensing nuclear exports. India, a non-party to the NPT, does not have full-scope safeguards, nor is it ever expected to adopt full- scope safeguards, since it has a nuclear weapons program that would preclude them. Also at issue is the requirement in Section 129 to stop exports if a non-nuclear-weapon state has detonated a nuclear device after 1978, among other things. India detonated several nuclear devices in 1998.

These three sections of the AEA provide mechanisms for the President to waive those requirements and sanctions (in Section 129), which are spelled out in more detail below. The sections also provide legislative vetoes, in the form of concurrent resolutions, of the presidential determinations. In 1983, however, the Supreme Court decided in *INS v. Chadha* that legislative veto provisions that do not satisfy the bicameralism and presentment requirements of Article I of the Constitution were unconstitutional. In 1985, some parts of the AEA were amended to provide for joint resolutions of approval or disapproval (e.g., Section 123 d.). The *Chadha* decision affects how Congress would disapprove of such presidential determinations under existing law and therefore affects the impact of the Administration's proposed legislation.

Agreements for Cooperation

Section 123 of the AEA (42 U.S.C. 2153) specifies what must happen before nuclear cooperation can take place:

- **Section 123 a.** states that the proposed agreement shall include the terms, conditions, duration, nature, and scope of cooperation and lists nine criteria that the agreement must meet. It also contains provisions for the President to exempt an agreement from any of the nine criteria, and includes details on the kinds of information the executive branch must provide to Congress;
- **Section 123 b.** specifies the process for submitting the text of the agreement to Congress;
- **Section 123 c.** specifies how Congress approves cooperation agreements that are limited in scope (e.g., do not transfer nuclear material or cover reactors larger than 5 MWe.);[130] and
- **Section 123 d.** specifies how Congress approves agreements that do cover significant nuclear cooperation (transfer of nuclear material or reactors larger than 5 MWe), including exempted agreements.

The United States has more than 20 agreements for cooperation in place now, and had an agreement with India from 1963 to 1993. Such agreements for cooperation are "framework" agreements—they do not guarantee that cooperation will take place or that nuclear material will be transferred, but rather set the terms of reference and provide authorization for cooperation. The 1963 U.S.-India cooperation agreement is anomalous in that it did guarantee fuel for the Tarapur reactors, even though other U.S. nuclear cooperation agreements reportedly have not included any such guarantees.[131]

Section 123 a. lists nine criteria that an agreement must meet unless the President determines an exemption is necessary. These are listed in Section 123 a., paragraphs (1) through (9), 42 U.S.C. 2153. They are guarantees that (1) safeguards on nuclear material and equipment transferred continue in perpetuity; (2) full-scope safeguards are applied in non-nuclear-weapon states; (3) nothing transferred is used for any nuclear explosive device or for any other military purpose; (4) the United States has the right of return if the cooperating state detonates a nuclear explosive device or terminates or abrogates an International Atomic Energy Agency (IAEA) safeguards agreement; (5) there is no transfer of material or classified data without U.S. consent; (6) physical security is maintained; (7) there is no enrichment or reprocessing by the recipient state without prior approval; (8) storage is approved by United States for plutonium and highly enriched uranium; and (9) anything produced through cooperation is subject to all of the above requirements.

In the case of India, the most difficult of these requirements to meet is the full-scope safeguards requirement for non-nuclear-weapon states (Sec. 123 a. (2)). India is considered to be a nonnuclear-weapon state because it did not, as defined by the nuclear Nonproliferation Treaty, explode a nuclear device before January 1, 1967.[132] The President may exempt an agreement for cooperation from any of the requirements in Section 123 a. if he determines that meeting the requirement would be "seriously prejudicial to the achievement of U.S. non-proliferation objectives or otherwise jeopardize the common defense and security." An exempted agreement would not become effective "unless the Congress adopts, and there is enacted, a joint resolution stating that the Congress does favor such agreement."[133] In other words, both chambers of Congress must approve the agreement if it does not contain all of the Section 123 a. requirements.

If Congress votes to approve an agreement for cooperation that was exempted because the recipient state did not have full-scope safeguards (Section 123 a. (2)), such approval would essentially waive the Nuclear Regulatory Commission's (NRC) obligation to consider full-scope safeguards as an export license authorization criterion under Section 128. However, Congress would still have the authority to review one export license authorization approximately every 12 months after the agreement for cooperation has entered into force (see discussion below).

Section 123 d., in part, states the following:

> if Congress fails to disapprove a proposed agreement for cooperation which exempts the recipient nation from the requirement set forth in subsection 123 a. (2), such failure to act shall constitute a failure to adopt a resolution of disapproval pursuant to subsection 128 b. (3) for purposes of the Commission's consideration of applications and requests under section 126 a. (2) and there shall be no congressional review pursuant to section 128 of any subsequent license or authorization with respect to that state until the first such license or authorization which is issued after twelve months from the elapse of the sixty-day period in which the agreement for cooperation in question is reviewed by the Congress.[134]

Export Licensing

In addition to specifying criteria for framework agreements, the AEA sets out procedures for licensing exports (Sections 126, 127, and 128 codified as amended at 42 U.S.C. 2155, 2156, 2157). The NRC is required to meet criteria in Sections 127 and 128 for authorizing export licenses; Section 128 contains the requirement for full-scope safeguards for non-nuclear-weapon states. Section 126 b. (2) contains a provision for the President to authorize an export in the event that the NRC deems that the export would not meet Section 127 and 128 criteria. The President must determine "that failure to approve an export would be seriously prejudicial to the achievement of U.S. nonproliferation objectives or otherwise jeopardize the common defense and security." In that case, the President would submit his executive order, along with a detailed assessment and other documentation, to Congress for 60 days of continuous session. After 60 days of continuous session, the export would go through unless Congress were to pass a concurrent resolution of disapproval.[135]

In the case of exports pursuant to an exempted agreement for cooperation as described above, the NRC does not have to meet the full-scope safeguards requirement in assessing whether it can issue export licenses (Section 128 b. (3)). If the NRC were to issue such a license, Congress would review one license every 12 months. If Congress were to pass a resolution of disapproval, no further exports could be made during that Congress.[136]

In both cases, Section 128 contains a provision for the President to waive termination of exports by notifying the Congress that the state has adopted full-scope safeguards or that the state has made significant progress toward full-scope safeguards, or that U.S. foreign policy interests dictate reconsideration. Such a determination would become effective unless Congress were to disagree with the President's determination.[137]

Termination of Cooperation

Section 129 of the AEA (42 U.S.C. 2158) requires ending exports of nuclear materials and equipment or sensitive nuclear technology to any non-nuclear-weapon state that, after March 10, 1978, the President determines to have

- detonated a nuclear explosive device;
- terminated or abrogated IAEA safeguards;
- materially violated an IAEA safeguards agreement; or
- engaged in activities involving source or special nuclear material and having "direct significance" for the manufacture or acquisition of nuclear explosive devices, and "has failed to take steps which, in the President's judgment, represent sufficient progress toward terminating such activities."

In addition, Section 129 would also halt exports to any nation the President determines to have

- materially violated the terms of an agreement for cooperation with the United States, or
- assisted, encouraged, or induced any other non-nuclear-weapon state to obtain nuclear explosives or the materials and technologies needed to manufacture them; or

re-transferred or entered into an agreement for exporting reprocessing equipment, materials or technology to another non-nuclear-weapons state.

The President can waive termination if he determines that "cessation of such exports would be seriously prejudicial to the achievement of United States nonproliferation objectives or otherwise jeopardize the common defense and security." The President must submit his determination to Congress, which is then referred to the House International Relations Committee and the Senate Foreign Relations Committee for 60 days of continuous session. The determination becomes effective unless Congress opposes it.[138]

P.L. 109-401[139]

On March 9, 2006, the Administration submitted its proposed legislation to Representative Hyde and Senator Lugar, and on March 16, 2006, Representatives Hyde and Lantos introduced H.R. 4974, and Senator Lugar introduced S. 2429. Following public hearings and committee mark-ups, the House passed H.R. 5682 on July 26, 2006, by a vote of 359 to 68 and the Senate passed its version of H.R. 5682, substituting the text of the amended S. 3709, on November 16, 2006, by a vote of 85 to 12. One issue that held up the Senate bill was the inclusion, in Title II, of the implementing legislation for the U.S. Additional Protocol—an agreement between the United States and the IAEA to provide for enhanced information, access, and inspection tools for IAEA inspectors as they inspect U.S. nuclear and other facilities under the U.S. voluntary safeguards agreement.

The House and Senate versions of H.R. 5682 were remarkably similar, with four differences.[140] The Senate version contained an additional requirement for the President to execute his waiver authority, an amendment introduced by Senator Harkin and adopted by unanimous consent that the President determine that India is "fully and actively participating in U.S. and international efforts to dissuade, sanction and contain Iran for its nuclear program." This provision was changed to a reporting requirement in the conference report. The Senate version also had two unique sections related to the cooperation agreement, Sections 106 and 107, both of which appear in the conference report. Section 106 (now Section 104 (d) (4)) prohibits exports of equipment, material or technology related for uranium enrichment, spent fuel reprocessing or heavy water production unless conducted in a multinational facility participating in a project approved by the IAEA or in a facility participating in a bilateral or multilateral project to develop a proliferation- resistant fuel cycle. Section 107 (now Section 104 (d) (5)) would establish a program to monitor that U.S. technology is being used appropriately by Indian recipients. Finally, the Senate version also contained the implementing legislation for the U.S. Additional Protocol in Title II, which was retained in the conference report.

P.L. 109-401 allows the President to

- exempt a proposed agreement for cooperation with India from the full-scope safeguards requirement of Section 123 a. (2) of the Atomic Energy Act;
- exempt an agreement from any export review by the Congress under Section 128 of the AEA; and

- exempt the agreement from restrictions resulting from India's nuclear weapons activities under Section 129 a. (1) (D) of the AEA, and exempt the agreement from a cutoff in exports because of India's 1998 nuclear test.

It does not exempt the agreement from a future cutoff in exports if India tests a nuclear explosive device again. For the President to exercise his waiver authority, seven requirements, as outlined earlier, must be met. P.L. 109-401 contains numerous statements of policy and reporting requirements, as well as restrictions on certain kinds of transfers. There are specific prohibitions on (as outlined in Section 104 (d)): (1) transfers that would violate U.S. obligations under Article 1 of the NPT not to in any way assist any country to manufacture or otherwise acquire nuclear weapons; (2) transfers that would violate NSG guidelines in force at the time; (3) enrichment and reprocessing cooperation, except to "a multinational facility participating in an IAEA-approved program to provide alternatives to national fuel cycle capabilities; or ... a facility participating in, and the export, reexport, transfer, or retransfer is associated with, a bilateral or multinational program to develop a proliferation-resistant fuel cycle." Additionally, the law requires a cutoff in exports if India is found to have violated NSG or MTCR guidelines. P.L. 109-401 also provides for a nuclear export accountability program (formerly Section 107 of the Senate version of H.R. 5682).

President's Signing Statement

In President Bush's signing statement, he noted that the act "will strengthen the strategic relationship between the United States and India."[141] With respect to particular provisions, President Bush stated that the executive branch would construe two sections of the bill as "advisory" only: policy statements in Section 103 and the restriction contained in Section 104 (d) (2) on transferring items to India that would not meet NSG guidelines. On the first, the President cited the Constitution's "commitment to the presidency of the authority to conduct the Nation's foreign affairs"; on the second, the President raised the question of whether the provision "unconstitutionally delegated legislative power to an international body." In other words, the President was questioning whether Congress were ceding authority to approve U.S. exports to the NSG. However, U.S. officials, including Secretary of State Rice, have formally told Congress multiple times that the United States government would abide by NSG guidelines. The President's signing statement also noted that the executive branch would construe

> provisions of the Act that mandate, regulate, or prohibit submission of information to the Congress, an international organization, or the public, such as sections 104, 109, 261, 271, 272, 273, 274, and 275, in a manner consistent with the President's constitutional authority to protect and control information that could impair foreign relations, national security, the deliberative processes of the Executive, or the performance of the Executive's constitutional duties.

This seems to suggest that the executive branch might limit the scope of reporting required by Congress in those sections, not just on national security grounds, but to protect executive branch processes or performance. The implications of the approach outlined in this signing statement will not be clear until the executive branch produces (or does not produce, as the case may be) required reports.

Issues for Congress

The Nuclear Cooperation Agreement

The agreement announced in July 2007 by the United States and India lists a variety of civilian nuclear projects on which the two countries "may pursue cooperation." Although the Bush Administration argued that the agreement "is consistent with applicable U.S. law,"[142] some Members of Congress expressed concern that it may be inconsistent with parts of P.L. 109-401. For example, H.Res. 711, which was referred to the House Committee on Foreign Affairs October 4, 2007, stated that "it is the sense of the House of Representatives" that the Bush Administration should not propose changes to NSG guidelines until it has resolved "all differences of interpretation" of the agreement with New Delhi and "answered all outstanding questions raised by Congress regarding apparent inconsistencies between the nuclear cooperation agreement" and P.L. 109-401. Non-governmental experts also raised questions about several aspects of the agreement, arguing that they may be inconsistent with P.L. 109-401.[143]

Sensitive Nuclear Technology Transfers

Despite restrictions in P.L. 109-40 1 regarding U.S. exports of equipment, material or technology related to uranium enrichment, spent fuel reprocessing or heavy water production, the agreement states that

> Sensitive nuclear technology, heavy water production technology, sensitive nuclear facilities, heavy water production facilities and major critical components of such facilities may be transferred under this Agreement pursuant to an amendment to this Agreement.

However, the agreement also states that "transfers of dual-use items that could be used in enrichment, reprocessing or heavy water production facilities will be subject to the Parties' respective applicable laws, regulations and license policies." Such transfers would, therefore, be subject to the same restrictions described in P.L. 109-401. Any other transfers of such technology would require changes to existing U.S. law. The State Department has said that Washington does not intend to negotiate an amendment to the agreement.[144] Furthermore, "as a matter of policy, the United States does not transfer dual-use items for use in sensitive nuclear facilities" and "will not assist India in the design, construction, or operation of sensitive nuclear technologies through the transfer of dual-use items, whether under the [123] Agreement or outside the Agreement," according to the State Department.[145]

Nuclear Testing/Right of Return

P.L. 109-401 does not exempt the agreement from a future cutoff in exports if India tests a nuclear explosive device again.[146] However, the agreement does not explicitly mention U.S. responses to such a test of such a device. Instead, the agreement states that "either Party shall have the right to terminate this Agreement prior to its expiration on one year's written notice to the other Party"—that is, the agreement does not limit the grounds upon which the agreement may be terminated.

Similarly, the agreement provides that the party seeking termination

> has the right to cease further cooperation under this Agreement if it determines that a mutually acceptable resolution of outstanding issues has not been possible or cannot be achieved through consultations.

This provision means that nuclear cooperation under the agreement may be terminated by a party during the one-year notice period for termination of the agreement.

The agreement also specifies that the two governments are to "hold consultations" prior to ceasing cooperation or terminating the agreement. The United States and India are to

> take into account whether the circumstances that may lead to termination or cessation resulted from a Party's serious concern about a changed security environment or as a response to similar actions by other States which could impact national security.

This provision suggests that, in the event that India conducts a nuclear explosive test, New Delhi may argue that the agreement should not be terminated (and nuclear cooperation should not cease) because geopolitical circumstances justified the test. However, in such cases, the U.S. right to terminate and cease cooperation under this provision would not be constrained by the results of the consultations.

With regard to the U.S. right of return, Section 123 a. (4) of the AEA requires that nuclear cooperation agreements include

> a stipulation that the United States shall have the right to require the return of any nuclear materials and equipment transferred pursuant thereto and any special nuclear material produced through the use thereof if the cooperating party detonates a nuclear explosive device or terminates or abrogates an agreement providing for IAEA safeguards.

The July agreement states that, following the cessation of cooperation under this agreement, either party has the right to require the return of "any nuclear material, equipment, non-nuclear material or components transferred under this Agreement and any special fissionable material produced through their use." However, the agreement does not say explicitly that a future Indian test of a nuclear explosive device would allow the United States to exercise its right of return. Rather, it provides for a right of return whenever a party has given notice of termination of the agreement and has ceased nuclear cooperation, which would include but not be limited to the circumstances specified in section 123 .a(4) of the AEA. The agreement also provides that a "notice by a Party that is invoking the right of return shall be delivered to the other Party on or before the date of termination of this Agreement." This means that the right of return cannot be exercised after the one-year interval prior to the agreement's termination.

Fuel Supply

A closely related issue is the agreement's four assurances regarding India's future nuclear fuel supply:

- The United States is willing to incorporate assurances regarding fuel supply in the bilateral U.S.-India agreement on peaceful uses of nuclear energy under Section 123 of the U.S. AEA, which would be submitted to the U.S. Congress.
- The United States will join India in seeking to negotiate with the IAEA an India-specific fuel supply agreement.
- The United States will support an Indian effort to develop a strategic reserve of nuclear fuel to guard against any disruption of supply over the lifetime of India's reactors.
- If, despite these [above] arrangements, a disruption of fuel supplies to India occurs, the two governments would jointly convene a group of friendly supplier countries (including countries such as Russia, France, and the United Kingdom) to pursue such measures as would restore fuel supply to India.

The last two provisions are particularly controversial because they could potentially provide India a way to mitigate the effects of a U.S. cessation of nuclear exports (in the event that, for example, India were to test a nuclear weapon). [147] Indeed, Indian Foreign Secretary Shyam Saran asserted in a February 18, 2008, statement that, under the 123 agreement, India is entitled to build a strategic fuel reserve "to last the lifetime of such reactors." And a spokesperson for India's ruling Congress Party indicated in September 2008 that reserve supplies of fuel would enable New Delhi to continue operating its reactors even if other countries were to halt cooperation in response to an Indian nuclear test. [148]

P.L. 109-401 contains several provisions that could be in tension with the July agreement. For example, Section 103 (b) (10) addresses the issue of a fuel reserve:

> Any nuclear power reactor fuel reserve provided to the Government of India for use in safeguarded civilian nuclear facilities should be commensurate with reasonable reactor operating requirements.

With regard to supplying India with nuclear fuel after a nuclear test, Section 103 a. (6)[149] says that the United States should

> Seek to prevent the transfer to a country of nuclear equipment, materials, or technology from other participating governments in the NSG or from any other source if nuclear transfers to that country are suspended or terminated.

Similarly, Section 102 (13) expresses the sense of Congress that the United States "should not seek to facilitate or encourage the continuation of nuclear exports to India by any other party if such exports are terminated under United States law."

However, President Bush's September 10 message transmitting the agreement to Congress characterizes the agreement's fuel-supply assurances as "political commitments" that are not "legally binding" because the agreement is only a "framework agreement" that does not compel specific nuclear cooperation. Furthermore, according to the State

Department, the "disruption of fuel supplies" referred to in the agreement refers only to disruptions "that may result through no fault" of India's.[150]

Regarding the question of non-U.S. suppliers, Washington has not sought commitments from any other country to supply fuel to India. The "United States would be compelled to encourage transfers of nuclear fuel to India by other" NSG members if supply disruptions "occur through no fault of India's own," according to the State Department.[151] However, these assurances "are not ... meant to insulate India against the consequences of a nuclear explosive test or a violation of nonproliferation commitments." Indeed, such U.S. commitments "would no longer apply" if the United States were to terminate the agreement in response to an Indian nuclear test.

With respect to fuel reserves, the agreement does not define what it means to "support an Indian effort to develop a strategic reserve." Furthermore, the State Department suggests that the United States may not supply India with a fuel reserve sufficient for the lifetime of India's reactors, though the department's 2008 responses to Questions for the Record do not specify the size of any such reserve.[152]

P.L. 110-369—KEY PROVISIONS

As noted above, the House passed H.R. 7081, which approved the nuclear cooperation agreement, September 27, 2008.[153] The Senate Committee on Foreign Relations had, following a September 18 hearing and subsequent markup, approved identical legislation, S. 3548 (introduced by Senator Christopher Dodd), September 23. The Senate passed H.R. 7081 October 1. President Bush signed P.L. 110-369 into law October 8, 2008.

P.L. 110-369, the United States-India Nuclear Cooperation Approval and Nonproliferation Enhancement Act, obviated the 30-day consultative period, as well as other procedures cited in sections 123 b. and d. of the AEA. Section 101 states that "notwithstanding the provisions for congressional consideration and approval of a proposed agreement for cooperation" in those two sections, "Congress hereby approves the United States-India Agreement for Cooperation on Peaceful Uses of Nuclear Energy." It also states that the agreement "shall be subject to" applicable U.S. law as if it had been approved according to section 123's provisions.

Declarations of Policy

P.L. 110-369 contains several declarations of U.S. policy. Section 102 (a) states that "it is the understanding of the United States" that the agreement's provisions

> have the meanings conveyed in the authoritative representations provided by the President and his representatives to the Congress and its committees prior to September 20, 2008, regarding the meaning and legal effect of the Agreement.

As noted above, some lawmakers had previously expressed concern about ambiguities in the agreement and whether it met the requirements of P.L. 109-401. During the debate over the bill, Berman stated September 26 that "I continue to have concerns about ambiguities in the agreement" and inserted the State Department's January 2008 responses to the committee's questions into the record in order to "clarify the meaning of these and other important issues." The documents "constitute key and dispositive parts of the 'authoritative representations' described in section 102," he added.

The law contains two other provisions apparently designed to clarify that the agreement's fuel reserve and fuel supply provisions are not intended to provide New Delhi a way to test nuclear weapons without fear of consequence. The first, section 102 (b) (1), states that "in the event that nuclear transfers to India are suspended or terminated" pursuant to U.S. law, "it is the policy of the United States to seek to prevent the transfer to India of nuclear equipment, materials, or technology" from other NSG participants "or from any other source." This provision is also contained in section 103 (a)(6) of P.L. 109-401.

The second, section 102 (b)(2), restates a provision contained in section 103 (b) (10) of P.L. 109- 401 regarding the fuel reserve:

> Any nuclear power reactor fuel reserve provided to the Government of India for use in safeguarded civilian nuclear facilities should be commensurate with reasonable reactor operating requirements.

Certification Requirements

P.L. 110-369 contains two certification requirements that had to be met before the United States could exchange diplomatic notes with India—a step which, as noted above, was necessary for the agreement to enter into force. Section 102 (c) requires the President to certify to Congress that the agreement is "consistent with" U.S. obligations under Article I of the NPT. Section 204 (a) requires the President to certify to the Senate Foreign Relations Committee and the House Foreign Affairs Committee that it is U.S. policy to work with NSG members "to agree to further restrict the transfers of equipment and technology related to the enrichment of uranium and reprocessing of spent nuclear fuel." President Bush transmitted the certifications, along with a Memorandum of Justification, October 20, 2008. The memorandum did not explain the reasoning underlying the determinations.

In addition, Section 204 (b) states that

> the President shall seek to achieve, by the earliest possible date, either within the NSG or with relevant NSG Participating Governments, the adoption of principles, reporting, and exchanges of information as may be appropriate to assure peaceful use and accounting of by-product material in a manner that is substantially equivalent to the relevant provisions [of the nuclear cooperation agreement].

There is no certification requirement for this provision, although section 204(c) requires the President to submit a report every six months on U.S. efforts to achieve these changes.

Section 104 of P.L. 110-369 requires that, before the NRC can issue export licenses, the President must determine and certify to Congress that India's IAEA safeguards agreement has entered into force and that New Delhi's declaration of its nuclear facilities to the IAEA "is not materially inconsistent with the facilities and schedule" described in India's separation plan.

As noted, India signed its safeguards agreement February 2, 2009, and it entered into force May 11, 2009. New Delhi has also, as noted, submitted its declarations of nuclear facilities pursuant to paragraphs 13 and 14 of its safeguards agreement. President Obama submitted the required certification to Congress February 3, 2010, determining that India has satisfied the legal requirement described above. The certification did not describe the reasoning underlying the determination.

Reporting Requirements

P.L. 110-369 adds several reporting requirements to P.L. 109-401. It amends section 104 (g)(1) to require that the President inform the Senate Foreign Relations Committee and the House Foreign Affairs Committee of "any material inconsistencies" with respect to content or timing between India's separation plan and the notifications New Delhi is to provide to the IAEA pursuant to paragraph 14 of India's safeguards agreement. P.L. 110-369 also amends section 104(g)(2) to require the President to report on a variety of activities that could be undertaken pursuant to the nuclear cooperation agreement.

Section 202 of P.L. 110-369 amends section 123 of the AEA to require the President to keep the Senate Foreign Relations Committee and the House Foreign Affairs Committee "fully and currently informed of any initiative or negotiations relating to a new or amended agreement for peaceful nuclear cooperation."[154]

Procedures for Subsequent Arrangements

As noted, the United States and India signed an agreement July 30, 2010, concerning the arrangements and procedures for India to reprocess spent nuclear fuel in a new reprocessing facility. Section 201 of P.L. 110-369 specifies procedures for Congress to consider such a subsequent arrangement. First, the President must transmit to the Senate Foreign Relations Committee and the House Foreign Affairs Committee a report describing the reasons for the proposed arrangement, a description (including the text) of the arrangement, and a certification that the United States

> will pursue efforts to ensure that any other nation that permits India to reprocess or otherwise alter in form or content nuclear material that the nation has transferred to India or nuclear material and by-product material used in or produced through the use of nuclear material, non-nuclear material, or equipment that it has transferred to India requires India to do so under similar arrangements and procedures.

In addition, 30 days of continuous session must elapse after the President has submitted the report. The proposed arrangement shall not take effect if Congress adopts a joint resolution of disapproval within this 30-day period. Section 201 requires that such a resolution "be considered pursuant to the procedures set forth in section 130 i" of the AEA. Section 205, however, shortens from 45 to 15 days the amount of time that the Senate Foreign Relations Committee and the House Foreign Affairs Committee have to report the resolution.[155] Secretary of Energy Steven Chu submitted the subsequent arrangement and related materials to Congress May 11, 2010.

APPENDIX. INDIA'S SEPTEMBER 5, 2008, STATEMENT ON DISARMAMENT AND NONPROLIFERATION

Statement by External Affairs Minister of India Shri Pranab Mukherjee on the Civil Nuclear Initiative

To reiterate India's stand on disarmament and nonproliferation, EAM has made the following statement:

> A Plenary meeting of the Nuclear Suppliers Group to consider an exception for India from its guidelines to allow for full civil nuclear cooperation with India is being held in Vienna from September 4-5, 2008.
>
> India has a long-standing and steadfast commitment to universal, non-discriminatory and total elimination of nuclear weapons. The vision of a world free of nuclear weapons which Shri Rajiv Gandhi put before the UN in 1988 still has universal resonance.
>
> We approach our dialogue with the Nuclear Suppliers Group and all its members in a spirit of cooperation that allows for an ongoing frank exchange of views on subjects of mutual interest and concern. Such a dialogue will strengthen our relationship in the years to come.
>
> Our civil nuclear initiative will strengthen the international non-proliferation regime. India believes that the opening of full civil nuclear cooperation will be good for India and for the world. It will have a profound positive impact on global energy security and international efforts to combat climate change.
>
> India has recently submitted a Working Paper on Nuclear Disarmament to the UN General Assembly, containing initiatives on nuclear disarmament. These include the reaffirmation of the unequivocal commitment of all nuclear weapon States to the goal of complete elimination of nuclear weapons; negotiation of a Convention on the complete prohibition of the use or threat of use of nuclear weapons; and negotiation of a Nuclear Weapons Convention prohibiting the development, production, stockpiling and use of nuclear weapons and on their destruction, leading to the global, non-discriminatory and verifiable elimination of nuclear weapons within a specified timeframe.
>
> We remain committed to a voluntary, unilateral moratorium on nuclear testing. We do not subscribe to any arms race, including a nuclear arms race. We have always tempered the exercise of our strategic autonomy with a sense of global responsibility. We affirm our policy of no-first-use of nuclear weapons.
>
> We are committed to work with others towards the conclusion of a multilateral Fissile Material Cut-off Treaty in the Conference on Disarmament that is universal, non-discriminatory and verifiable.
>
> India has an impeccable non-proliferation record. We have in place an effective and comprehensive system of national export controls, which has been constantly updated to meet the highest international standards. This is manifested in the enactment of the Weapons of Mass Destruction and their Delivery Systems Act in 2005. India has taken the necessary steps to secure nuclear materials and technology through comprehensive export control legislation and through harmonization and committing to adhere to Missile Technology Control Regime and Nuclear Suppliers Group guidelines.

India will not be the source of proliferation of sensitive technologies, including enrichment and reprocessing transfers. We stand for the strengthening of the non-proliferation regime. We support international efforts to limit the spread of ENR equipment or technologies to states that do not have them. We will work together with the international community to advance our common objective of non-proliferation. In this regard, India is interested in participating as a supplier nation, particularly for Thorium-based fuel and in establishment of international fuel banks, which also benefit India.

India places great value on the role played by the IAEA's nuclear safeguards system. We look forward to working with the IAEA in implementing the India-specific Safeguards Agreement concluded with the IAEA. In keeping with our commitment to sign and adhere to an Additional Protocol with respect to India's civil nuclear facilities, we are working closely with the IAEA to ensure early conclusion of an Additional Protocol to the Safeguards Agreement.

End Notes

[1] The full text of the agreement, which was released August 3, 2007, can be found at http://www.state 2007/aug/90050.htm.

[2] The relevant documents are available at http://foreignaffairs.house.gov/press_display.asp?id=555.

[3] Text Available at http://svaradarajan.blogspot.com/2008/10/dear-bill-foreign-secretarys-letter.html.

[4] Amol Sharma, "India Weighs Measure to Ease Nuclear Liability," *The Wall Street Journal*, October 10, 2010.

[5] "Questions for the Record Submitted to Under Secretary William Burns and Acting Under Secretary John Rood by Senator Robert P. Casey, Senate Foreign Relations Committee, September 18, 2008."

[6] The Lok Sabha adopted the legislation August 25; the Rajya Sabha adopted it August 30.

[7] Natalie Obiko Pearson, "India Risks Nuclear Isolation with Break From Chernobyl Accord," *Bloomberg*, August 26, 2010; Ann MacLachlan, Yanmei Xie, "India Liability Bill Seen Shaking Up Nuclear Trade, Liability Regime," *Nucleonics Week*, September 2, 2010.

[8] Ann MacLachlan, "Indian Regulator Says Country's Liability Law Compatible with CSC," *Nucleonics Week*, September 23, 2010.

[9] Analyst interview with U.S. officials, November 3, 2009.

[10] The text is available at http://www.state

[11] Rama Lakshmi, "Despite Historic Pact, U.S. Firms Are Hampered in Setting Up Reactors in India," *The Washington Post*, January 21, 2009. T.S. Subramanian, "India Atomic Body Chief: 123 Agreement Gives Spent Fuel, Reprocessing Rights," *The Hindu*, August 3, 2009.

[12] This report discusses only nuclear cooperation agreements that India has signed or otherwise concluded with other countries. India has discussed other nuclear cooperation agreements that are not discussed in this section.

[13] "Indo-French Pact Not to Cover Nuclear Enrichment Technology Sale - Envoy," *Press Trust of India*, September 19, 2008.

[14] "France Willing to Reprocess Uranium for India," *The Hindu Business Line*, January 31, 2009.

[15] "Russian president amends 1992 decree so as to ease nuclear exports to India," *Interfax*, February 24, 2009.

[16] "Russia Delivers First Batch of Nuclear Fuel to India," *Press Trust of India*, April 10, 2009.

[17] "Kazakhstan signs agreement to supply uranium," *The Hindu*, January 25, 2009; Kazakhstan might start uranium exports to India in 2009," *Panorama*, February 6, 2009. "Chennai Daily Report: India, Kazakhstan Set To Sign Nuclear Reactor Export Deal," *Chennai Business Line Online*, July 10, 2009.

[18] "India Signs Civil Nuclear Pact with Mongolia," *Press Trust of India*, September 14, 2009.

[19] "UK, India Sign Civil Nuclear Accord, Paves Way for Conference," *Reuters*, February 13, 2010.

[20] http://www.thehindu.com/news/national/article490238.ece

[21] For an analysis of the proliferation implications of U.S. nuclear exports to India, see Gary Milhollin, "Stopping the Indian Bomb," *The American Journal of International Law*, July 1987, 81 A.J.I.L. 593. See http://www.wisconsinproject.org/pubs/articles/1987/stoppingindianbomb.htm.

[22] The NNPA, in part, amended the Atomic Energy Act of 1954. See 42 U.S.C. 2151 et seq. Prior to the 1970 NPT, safeguards (inspections, material protection, control and accounting) were applied to specific facilities or materials (known as INFCIRC/66-type agreements). The NPT required safeguards on all nuclear material in all peaceful nuclear activities for non-nuclear-weapon-state parties (those states not having detonated a nuclear explosive device prior to January 1, 1967).

[23] IAEA Document INFCIRC/254, *Guidelines for Transfers of Nuclear-related Dual-use Equipment, Materials, Software, and Related Technology*. Part 1 covers "trigger list" items: those especially designed or prepared for nuclear use: (i) nuclear material; (ii) nuclear reactors and equipment; (iii) non-nuclear material for reactors; (iv) plant and equipment for reprocessing, enrichment and conversion of nuclear material and for fuel fabrication and heavy water production; and (v) associated technology. Part 2 covers dual-use items. Additional NSG criteria for dual-use exports include NPT membership and/or full-scope safeguards agreement; appropriate end-use; whether the technology would be used in a reprocessing or enrichment facility; the state's support for nonproliferation; and the risk of potential nuclear terrorism.

[24] China was not a member of the NSG until 2004. Russia, an NSG member, exported fuel, citing a safety exception, but NSG members objected so strongly that Moscow suspended supply in 2004. Russia agreed to resupply Tarapur in late February and informed the NSG on February 27, 2006, reportedly citing the NSG safety exception.

[25] See also CRS Report RL33072, *U.S.-India Bilateral Agreements and "Global Partnership"*, by K. Alan Kronstadt, and CRS Report RL33529, *India-U.S. Relations*, by K. Alan Kronstadt.

[26] See fact sheet on the NSSP at http://www.state

[27] Joint Statement Between President George W. Bush and Prime Minister Manmohan Singh, White House Press Release, July 18, 2005, Washington, DC (hereafter cited as "July 18 Joint Statement") http://www.whitehouse.gov/ news/releases/2005/07/2005071 8-6.html.

[28] July 18 Joint Statement.

[29] See CRS Report RS22474, *Banning Fissile Material Production for Nuclear Weapons: Prospects for a Treaty (FMCT)*, by Sharon Squassoni, for more detailed information about the issue and negotiations.

[30] Opening Statement, Chairman Richard G. Lugar, Senate Foreign Relations Committee hearing on "Implications of U.S.-India Nuclear Energy Cooperation," November 2, 2005 (hereafter referred to as November 2, 2005 SFRC India hearing).

[31] The House International Relations Committee held the following hearings: "The U.S. and India: An Emerging Entente?" (September 8, 2005); "The U.S.-India Global Partnership: The Impact on Nonproliferation" (October 26, 2005); and "U.S.-India Global Partnership: How Significant for American Interests?" (November 16, 2005); "The U.S.- India Global Partnership" (April 5, 2006); "U.S.-India Global Partnership: Legislative Options" (May 11, 2006). See http://www.internationalrelations.house.gov/ for testimonies of witnesses. The Senate Foreign Relations Committee held the following hearings: "Implications of U.S.-India Nuclear Energy Cooperation" (November 2, 2005); "U.S.- India Atomic Energy Cooperation: The Indian Separation Plan and the Administration's Legislative Proposal" (April 5, 2006); and "U.S.-India Atomic Energy Cooperation: Strategic and Nonproliferation Implications" (April 26, 2006). See http://foreign.senate.gov/hearing.html for testimonies.

[32] Statement of Under Secretary of State for Political Affairs, R. Nicholas Burns, September 8, 2005, House Committee on International Relations, Hearing on "The U.S. and India: An Emerging Entente?" (hereafter cited as "September 8, 2005, HIRC US-India hearing") p. 1.

[33] Remarks by Congressman Jim Leach, September 8, 2005, House International Relations Committee U.S.-India Hearing.

[34] Henry Sokolski, "Implementing the Indian Nuclear Deal: What's at Risk, What Congress Should Require," Briefing to Congress, September 2005.

[35] See "India's Growth Target Unrealistic," *Financial Times*, January 23, 2003, which quotes the Asia Development Bank.

[36] Brahma Chellaney, "US Deal is a Bad Choice for Power Generation," *International Herald Tribune*, December 27, 2005.

[37] Statement of Under Secretary of State for Political Affairs, R. Nicholas Burns, November 2, 2005, Senate Foreign Relations Committee Hearing on Implications of U.S.-India Nuclear Energy Cooperation.

[38] See Miriam Rajkumar, "Indian Independence," Carnegie Analysis, September 20, 2005, at http://www.carnegieendowment.org/npp/publications/index.cfm?fa=view&id=17486.

[39] Dr. Francine Frankel, Statement before the House International Relations Committee, November 16, 2005, "India's Potential Importance for Vital U.S. Geopolitical Objectives in Asia: A Hedge Against a Rising China?"

[40] "Press Briefing by Foreign Secretary on the events in UN and IAEA," New Delhi, September 26, 2005, available at http://www.indianembassy.org/press_release/2005/Sept/29.htm.

[41] Briefing by MEA Official Spokesperson on Draft Resolution on Iran in IAEA, New Delhi, September 24, 2005, available at http://www.indianembassy.org/press_release/2005/Sept/16.htm.

[42] See CRS Report RL34544, *Iran's Nuclear Program: Status*, by Paul K. Kerr.

[43] "U.S.-India Nuclear Deal Could Die, Envoy Warns," *Washington Post*, January 26, 2006.

[44] "Ahmadinejad Thanks India for Positive Stands on Iran in IAEA," IRNA, September 23, 2005.

[45] September 26, 2005 press briefing, op. cit.

[46] Iran Republic News Agency, "118 countries back Iran's nuclear program" *Iran Times*, September 18, 2006.

[47] See CRS Report RS22530, *India and Iran: WMD Proliferation Activities*, by Sharon Squassoni, for more information related to sanctions imposed for Indian transfers to Iran.

[48] See http://www.nti.org/e_research/profiles/Iran

[49] "India Denies Nuclear Cooperation with Iran," *Agence France Presse*, December 13, 2003.

[50] John Larkin and Jay Solomon, "As Ties Between India and Iran Rise, U.S. Grows Edgy," *Wall Street Journal*, March 24, 2005.

[51] Thionyl chloride is a Schedule 3 chemical under the Chemical Weapons Convention. It has military and civilian uses, and is widely used in the laboratory and in industry.

[52] See list of sanctions at http://www.state.

[53] The full text of Mukherjee's statement is in the Appendix.

[54] David Albright and Susan Basu, "India's Gas Centrifuge Program: Stopping Illicit Procurement and the Leakage of Technical Centrifuge Know-How." Available at http://www.isis-online.org/publications/southasia/indianprocurement.pdf.

[55] David Albright and Paul Brannan, "Indian Nuclear Export Controls and Information Security: Important Questions Remain." Available at http://www.isis-online.org/publications/southasia/India

[56] "Questions for the Record Submitted to Under Secretaries Nicholas Burns and Robert Joseph by Chairman Richard G. Lugar (#6), Senate Foreign Relations Committee, November 2, 2005."

[57] September 8, 2005, HIRC US-India hearing.

[58] See, for example, "Nuclear Nonproliferation System is Challenged," *People's Daily,* March 16, 2006.

[59] Available at http://verificationthoughts.blogspot.com/2008/07/indian-separation-plan.html.

[60] "Concern over Russian Plan to Sell Nuclear Reactor Fuel," *Financial Times*, March 15, 2006.

[61] David S. Jonas, "Variations on Non-nuclear: May the 'Final Four' Join the Nuclear Nonproliferation Treaty as Nonnuclear Weapon States While Retaining Their Nuclear Weapons?" *Michigan State Law Review,* Summer 2005, p. 417 ff. Mr. Jonas is General Counsel of the National Nuclear Security Agency.

[62] Ambassador Norman Wulf, "Observations from the 2000 NPT Review Conference," *Arms Control Today*, November 2000.

[63] September 8, 2005, HIRC US-India hearing.

[64] Fred McGoldrick, Harold Bengelsdorf, Lawrence Scheinman, "The U.S.-India Nuclear Deal: Taking Stock," *Arms Control Today*, October 2005, pp. 6-12. See http://www.armscontrol.org/act/2005_10/OCT-Cover.asp.

[65] Statement by Robert J. Einhorn, Center for Strategic and International Studies, "The U.S.-India Global Partnership: The Impact on Nonproliferation" October 26, 2005.

[66] "Backgrounder on India-U.S. Civilian Nuclear Energy Cooperation," Indian Embassy, July 29, 2005. See http://www.indianembassy.org/press_release/2005/July/29.htm.

[67] Prime Minister Singh, "Excerpts from Prime Minister's Reply to Discussion in Raja Sabha on Civil Nuclear Energy Cooperation with the United States." Remarks are available at the Indian Ministry of External Affairs website, http://mea.gov.in.

[68] "IAEA Director General Reacts to U.S.-India Cooperation Agreement," See http://www.iaea.org/NewsCenter/PressReleases/2005/prn200504.html. Critics of the IAEA point out that it is an organization that measures its success in part by how much nuclear material and how many facilities are under inspection.

[69] On September 29, 2004, the State Department published Public Notice 4845 in the Federal Register imposing sanctions pursuant to the Iran Nonproliferation Act of 2000. Two Indian scientists were named—Dr. Prasad and C. Surendar. The State Department has not revealed what technology or equipment was transferred, but both scientists have worked for the Nuclear Power Corporation of India, Ltd., a government-owned entity that runs India's nuclear power plants. The Indian embassy reported in December 2005 that sanctions on Dr. Surendar had been removed. See http://www.indianembassy.org/press_release/5.asp. In the December 30, 2005 Federal Register, Public Notice 5257 stated simply that sanctions on an Indian entity issued in Public Notice 4845 had been rescinded.

[70] September 8, 2005, HIRC US-India hearing.

[71] "Backgrounder on India-U.S. Nuclear Energy Cooperation," July 29, 2005.

[72] "Excerpts from Prime Minister's Reply," August 17, 2006, op. cit.

[73] See Zia Mian and M.V. Ramana, "Wrong Ends, Means, and Needs: Behind the U.S. Nuclear Deal with India," *Arms Control Today*, January/February 2006. See also Robert Einhorn, "Limiting the Damage," *The National Interest,* Winter 2005/2006.

[74] Statement of David Albright before the House International Relations Committee on October 26, 2005 (hereafter cited as "HIRC October 26, 2005 hearing").

[75] Questions for the Record Submitted to Secretary of State Condoleezza Rice by Senator Richard Lugar (#2) Senate Foreign Relations Committee, April 5, 2006.

[76] White House, Office of the Press Secretary, "Press Briefing by Under Secretary of State for Political Affairs Nick Burns," Maurya Sheraton Hotel and Towers, New Delhi, India, March 2, 2006.

[77] Mohamed ElBaradei, "Rethinking Nuclear Safeguards," *Washington Post*, June 14, 2006.

[78] See Henry Sokolski, "Fissile isn't Facile," *Wall Street Journal,* February 21, 2006.

[79] See Charles Ferguson, "Reshaping the U.S.-Indian Nuclear Deal to Lessen the Nonproliferation Losses," *Arms Control Today*, April 2008. Ferguson cites information from the Nuclear Power Corporation of India. See also "Uranium Shortage Hits Nuclear Power," *The Hindu Business Line*, May 21, 2008.

[80] Several countries have supplied low-enriched uranium to the U.S.-origin Tarapur reactors, including France, China and Russia.

[81] "India and the Nuclear Domain." Available at http://mea.gov.in/.

[82] Zia Mian, A.H. Nayyar, R. Rajaraman and M.V. Ramana, *Fissile Materials in South Asia: The Implications of the U.S.-India Nuclear Deal*, September 2006. Available at http://www.fissilematerials.org/ipfm/site_down/rr01.pdf.

[83] K. Subrahmanyam, former head of the Institute for Defence Studies and Analysis, was appointed Head of the National Security Council Advisory Board (NSCAB) established by the first Vajpayee government to draft the Indian nuclear doctrine. He currently chairs PM Singh's Global Strategic Developments Task Force. See also Dr. A. Gopalakrishnan, "Civilian and Strategic Nuclear Facilities of India," January 5, 2006.

[84] It is worth noting that even before the NPT entered into force, negotiators recognized that a state outside the treaty could preserve its domestic uranium sources for a possible weapons program as long as it agreed to accept IAEA safeguards on the items it imported. In the late 1960s, however, Congress was more concerned about ensuring that the United States could supply its allies outside the treaty, such as Japan and Germany, with nuclear fuel. According to Mason Willrich's history of the NPT, As long as India does not become a party to the Non-Proliferation Treaty, it can continue to import from the parties nuclear materials and equipment subject to safeguards for use in its civil nuclear power program. This would free its indigenous resources, particularly its limited uranium supply, for possible concentration on a nuclear weapons program. (Mason Willrich, *Non-proliferation Treaty: Framework for Nuclear Arms Control*, The Michie Company, Charlottesville, VA, 1969, p. 125.)

[85] "India Won't Define Minimum N-Deterrent: Saran," *Press Trust of India*, April 8, 2006.

[86] *Report Pursuant to Section 104 (c) of the Hyde Act Regarding Civil Nuclear Cooperation with India*. Submitted to Congress September 10, 2008. (Hereafter cited as, "2008 State Department Report.")

[87] See CRS Report RL3 3292, *India's Nuclear Separation Plan: Issues and Views*, by Sharon Squassoni, for details on the separation plan. See the previous section for additional issues concerning the separation plan.

[88] Prime Minister Singh presented "Implementation of the India-United States Joint Statement of July 18, 2005: India's Separation Plan," to Parliament on March 7, 2006. This is available at http://indianembassy.org/newsite/press_release/ 2006/Mar/sepplan.pdf. The plan was updated on May 11, 2006 to include names of reactors and upstream facilities, as well as dates they would be submitted to safeguards.

[89]According to the May 11 update, the eight indigenous reactors to be safeguarded are four at Rajasthan (RAPS 3, 4, 5 & 6); two at Uttar Pradesh (NAPS 1, 2); and two at Gujrat (KAPS 1, 2).

[90] Although India maintained a certain ambiguity by calling its 1974 test a "peaceful nuclear explosion," the 1998 tests leave little doubt that the experience gained was put to use in a nuclear weapons program. Plutonium produced in the CIRUS reactor, which the United States supplied with heavy water, was used in the 1974 test. See Victor Gilinsky and Paul Leventhal, "India Cheated," *Washington Post*, June 15, 1998. U.S. documents from the early 1970s indicate that the United States interpreted its nuclear cooperation agreement with India to prohibit peaceful nuclear explosions. Washington communicated this position to New Delhi. The documents are available at http://www.armscontrol.org/ country/india/Historic_Documents_India_Nuclear_Test.asp.

[91] The Additional Protocol is a measure to strengthen safeguards by providing for additional information, access and inspection tools. INFCIRC/540, concluded in 1997, is the model upon which states' protocols to their safeguards agreements are based.

[92] Statement of Dr. Joseph, November 2, 2005, SFRC India hearing.

[93] Ibid.

[94] Condoleezza Rice, "Our Opportunity With India," *Washington Post*, March 13, 2006.

[95] National Strategy to Combat Weapons of Mass Destruction, December 2002. Available at http://www.whitehouse.gov/news/releases/2002/12/WMDStrategy.pdf.

[96] See also CRS Report RL33529, *India-U.S. Relations*, by K. Alan Kronstadt.

[97] Those parties argued that the agreement would compromise India's sovereignty by drawing New Delhi into a "strategic alliance" with Washington. See, for example, http://www.cpim.org/ for a detailed account of objections from the Communist Party of India (Marxist).

[98] "N-deal strategic to India's progress: Pranab Mukherjee," *Indo Asian News Service*, February 3, 2008.

[99] "After Left Nod, Govt Decides To Go Ahead With IAEA Talks On N-Deal," *Financial Express*, November 16, 2007.

[100] Available at http://pmindia.nic.in/IaeaIndiaSGADrft.pdf.

[101] Mark Hibbs and Daniel Horner, "Pakistan, Other States Not Satisfied After Indian Briefing on IAEA Pact," *Nucleonics Week*, July 24, 2008.

[102] "Will Support Govt That Scraps 123," *The Economic Times*, September 8, 2008.

[103] Interview with Leader of the Opposition (Lok Sabha) L.K. Advani, *The Hindu*, July 11, 2008; "BJP Will Renegotiate N-Deal If Comes To Power: Sushma Swaraj," *The Economic Times*, September 7, 2008.

[104] Unless otherwise noted, this section is based on ElBaradei's August 1, 2008 comments to the IAEA Board of Governors, as well as personal communications with IAEA officials, and current and former State Department officials.

[105] Available at http://www.iaea.org/Publications/Documents/Infcircs/2008/infcirc731.pdf.

[106] Author interview, September 23, 2008.

[107] http://www.iaea.org/Publications/Documents/Infcircs/2009/infcirc754a1.pdf.

[108] Author interview with Indian official, January 11, 2010.

[109] "Indian Daily Interviews Atomic Body Chairman on Nuclear Deal," *The Hindu*, July 20, 2008.

[110] Ibid.

[111] Questions for the Record Submitted to Jeffrey Bergner, Assistant Secretary for Legislative Affairs by Rep. Tom Lantos, House Committee on Foreign Affairs (#12), October 5, 2007. Available at http://www.hcfa.house.gov/110/ press090208.pdf. (Hereafter, "Questions for the Record, 2007.")

[112] Ibid. (#11).

[113] Ibid. (#42).

[114] The Consultative Group is the NSG's "standing intersessional working body." See http://www.nuclearsuppliersgroup.org/activities.htm.

[115] A copy of the proposal is available at http://www.armscontrol.org/projects/india/20060327_DraftNSGProposal.asp.

[116] Author interviews with State Department official, February 4, 2008; August 14, 2008.

[117] The August 2008 draft is available at http://www.armscontrol.org/node/3274.

[118] The text of the NSG statement is available at http://www.nuclearsuppliersgroup.org/PRESS/2008-09-Press-Vienna.pdf.

[119] Austria, Ireland, the Netherlands, New Zealand, Norway, and Switzerland issued a joint statement at the August meeting explaining that they had proposed amendments to the U.S. proposal in order to meet their "non-proliferation objectives." New Zealand's Disarmament and Arms Control Minister Phil Goff stated August 26 that approximately 50 amendments had been proposed to the U.S.-proposed text.

[120] Mark Hibbs and Daniel Horner, "Scope of NSG Exemption For India Yet to Be Defined by Member States," *Nucleonics Week*, September 11, 2008.

[121] Wade Boese, "Nuclear Export Criteria Lacks Consensus," *Arms Control Today*, June 2008.

[122] See the Appendix.

[123] Rama Lakshmi, "U.S. Letter Puts India's Premier On Defensive Over Nuclear Deal," *Washington Post*, September 5, 2008.

[124] Author interview, September 23, 2008.

[125] State Department Report 2008.

[126] Ibid.

[127] Author interviews with State Department official August 11, 2008; August 14, 2008.

[128] Nuclear cooperation includes the distribution of special nuclear material, source material, and byproduct material, to licensing for commercial, medical, and industrial purposes. These terms, "special nuclear material," "source material," and "byproduct material," as well as other terms used in the statute, are defined in 42 U.S.C. § 2014.

[129] P.L. 83-703, 42 U.S.C. §§ 2153 et seq.

[130] In the 1954 Act, the provisions in Section 123 c. covered all agreements for cooperation. Section 123 d. was added in 1958 (P.L. 85-479) to cover military-related agreements. In 1974, P.L. 93-485 amended Section 123 d. to include agreements that covered reactors producing more than 5 MW thermal or special nuclear material connected therewith.

[131] United States General Accounting Office, "Nuclear Agreement: Cooperation Between the United States and the People's Republic of China," GAO/NSIAD-86-21BR, November 1985, Appendix I-1.

[132] 42 U.S.C. 2153 a.(2). Section 4 (b) of the NNPA specifies that all other terms used in the NNPA not defined in Section 4 "shall have the meanings ascribed to them by the 1954 Act, the Energy Reorganization Act of 1974 and the Treaty [NPT]." S.Rept. 95-467 further clarified that under the NPT, the five nuclear weapon states are the U.S., U.K., China, the Soviet Union, and France. U.S. Code Congressional and Administration News, 95th Cong., 2nd sess., 1978, vol. 3, p. 329.

[133] This new requirement was added by the Export Administration Amendments Act of 1985, P.L. 99-64, Section 301 (b) (2), 99 Stat. 120.

[134] The language "fails to disapprove" is an artifact of the 1978 Nuclear Nonproliferation Act, which used legislative vetoes in the form of concurrent resolutions of disapproval. In 1985, following the Supreme Court's *Chadha* decision invalidating the use of legislative vetoes, the Export Administration Amendments Act created a separate approval process for exempted agreements, which this part of Section 123 d. is referring to, that called for a joint resolution of approval. Thus, "fails to disapprove" could be interpreted as "approves" in the form of a joint resolution of approval.

[135] In light of the *Chadha* decision, passing a concurrent resolution could invite a legal challenge. Although this is not provided for in the AEA, Congress could choose to pass a joint resolution of disapproval or a bill stating in substance it did not approve.

[136] Section 128 b. (3) refers to a "resolution of disapproval," and this would likely be a joint resolution of disapproval, in light of the *Chadha* decision.

[137] Section 128 b. (2) refers to a "concurrent resolution." In light of the *Chadha* decision, Congress could pass a joint resolution disagreeing with the President's determination, or pass a bill barring nuclear exports for a certain period of time to that country.

[138] Section 129 specified that the President's determination "shall not become effective if during such sixty-day period the Congress adopts a concurrent resolution stating in substance that it does not favor the determination." However, P.L. 110-369 amended that section to change the "concurrent resolution" provision to a "joint resolution," presumably to overcome potential issues raised by the *Chadha* decision..

[139] See CRS Report RL33561, *U.S.-India Nuclear Cooperation: A Side-By-Side Comparison of Current Legislation*, by Sharon Squassoni and Jill Marie Parillo for more detail on the bills.

[140] See H.Rept. 109-590 and for S.Rept. 109-28 8, dated July 20, 2006, background on the bill.

[141] See http://www.whitehouse.gov/news/releases/2006/12/20061218-12.html.

[142] Department of State Fact Sheet July 27, 2007, at http://www.state

[143] See, for example, Sharon Squassoni, "Issues in U.S.-India Nuclear Cooperation," November 7, 2007 http://www.carnegieendowment.org/publications/index.cfm?fa=view&id=19697&prog=zgp&proj=znpp,zsa, and Daryl G. Kimball and Fred McGoldrick "U.S.-Indian Nuclear Agreement: A Bad Deal Gets Worse," August 3, 2007, at http://www.armscontrol.org/pressroom/2007/20070803_IndiaUS.asp?print.

[144] Questions for the Record, 2007 (#8).

[145] Ibid. (#s 4 and 5).

[146] As noted above, the President retains the right to waive the termination of nuclear exports.

[147] In a February 18, 2008 statement, Saran asserted that, under the 123 agreement, India is entitled to build a strategic fuel reserve "to last the lifetime of such reactors."

[148] "BJP, Left On 'Unholy Alliance' To Oppose N-Deal: Congress," *Press Trust of India*, September 5, 2008.

[149] As noted above, the President's signing statement says that he "shall construe" Section 103 as "advisory."

[150] Questions for the Record, 2007 (#16).

[151] Ibid. (#18).

[152] Ibid. (#19-20).

[153] H.R. 7039, a bill identical to H.R. 7081, was introduced by Representative Alana Ros-Lehtinen September 24. The next day, Representative Howard Berman introduced H.R. 7061, a bill very similar to H.R. 7081.

[154] H.R. 7061 required the President to consult with the House Foreign Affairs Committee and the Senate Foreign Relations Committee regarding any such initiative or negotiations.

[155] See CRS Report RS22937, *Nuclear Cooperation with Other Countries: A Primer*, by Paul K. Kerr and Mary Beth Nikitin.

In: U.S. Nuclear Cooperation with Other Countries
Editor: Gabriel A. Oliveira
ISBN: 978-1-61209-910-1

Chapter 6

U.S. AND SOUTH KOREAN COOPERATION IN THE WORLD NUCLEAR ENERGY MARKET: MAJOR POLICY CONSIDERATIONS

Mark Holt

SUMMARY

A South Korean consortium recently signed a contract to provide four commercial nuclear reactors to the United Arab Emirates (UAE), signaling a new role for South Korea in the world nuclear energy market. The $20 billion deal indicates that South Korea has completed the transition from passive purchaser of turn-key nuclear plants in the 1 970s to major nuclear technology supplier, capable of competing with the largest and most experienced nuclear technology companies in the world. The South Korean government reportedly has established a goal for South Korea to capture 20% of the world nuclear power plant market during the next 20 years, and the importance placed by Seoul on the UAE contract was underscored by South Korean President Lee Myung-bak's presence at the signing ceremony in the UAE.

In the 1970s, South Korea launched its nuclear power program through the government-owned Korea Electric Company (now Korea Electric Power Corporation, KEPCO), which purchased the country's first nuclear power units from Westinghouse. In the early years of the Korean nuclear program, Westinghouse and other foreign suppliers delivered completed plants with minimal Korean industry input. After the first three units, Korean firms took over the construction work on subsequent plants, although the reactor systems, turbine-generators, and architect/engineering services continued to be provided primarily by non-Korean companies. In 1987, KEPCO embarked on an effort to establish a standard Korean design, selecting the System 80 design from the U.S. firm Combustion Engineering as the basis. Combustion Engineering won the competition for the Korean standard design contract by agreeing to full technology transfer, according to KEPCO. The technology transfer program resulted in the development of the APR- 1400 power plant, which is the design purchased by the UAE.

In the UAE deal, the South Korean consortium is headed by KEPCO and includes other major Korean industrial companies that are involved in Korea's rapidly growing domestic nuclear power plant construction program. The consortium also includes Pittsburgh-based Westinghouse Electric Company, which currently owns the U.S. design on which the Korean design is based, and the Japanese industrial conglomerate Toshiba, which now owns most of Westinghouse.

Because the AP-1400 is based on a U.S. design, U.S. export controls will continue to apply. Westinghouse plans to seek the necessary authorization from the U.S. Department of Energy (DOE) to transfer the technology to the UAE. The UAE recently reached a nuclear cooperation agreement with the United States in which the country agreed not to develop fuel cycle facilities to support its planned nuclear power program, which could ease weapons proliferation concerns. The UAE program may establish a precedent for U.S. policy on future Korean exports to nonnuclear power nations, which is likely to be of continuing congressional interest.

The Korea-UAE nuclear plant sale also has been cited by the Korean news media as an important consideration in upcoming negotiations on the renewal of the U.S.-Korea nuclear cooperation agreement, a prerequisite under Section 123 of the Atomic Energy Act for nuclear trade. The current agreement expires in 2014, and the first discussions on renewal are likely within the next year. Congress will have an opportunity to review any new agreement before it takes effect.

INTRODUCTION: SOUTH KOREA'S NEW WORLD ROLE

South Korean consortium recently signed a contract to provide four commercial nuclear reactors to the United Arab Emirates (UAE), signaling a new role for South Korea in the world nuclear energy market. The $20 billion deal indicates that South Korea (the Republic of Korea, ROK) has completed the transition from passive purchaser of turn-key nuclear plants in the 1 970s to major nuclear technology supplier, now capable of competing with the largest and most experienced nuclear technology companies in the world.

Because the plants being exported by South Korea are based on a U.S. design, U.S. export controls will continue to apply. Westinghouse plans to seek the necessary authorization from the U.S. Department of Energy (DOE) to transfer the technology to the UAE. The UAE recently reached a nuclear cooperation agreement with the United States in which the country agreed not to develop fuel cycle facilities to support its planned nuclear power program, which could ease weapons proliferation concerns. The UAE program may establish a precedent for U.S. policy on future Korean exports to non-nuclear power nations, which is likely to be of continuing congressional interest.

The Korea-UAE nuclear plant sale has been cited by the Korean news media as an important consideration in upcoming negotiations on the renewal of the U.S.-Korea nuclear cooperation agreement, a prerequisite under Section 123 of the Atomic Energy Act for nuclear trade. Korean newspaper editorials have suggested that Korea's first foreign sale of nuclear technology could pave the way for U.S. agreement for Korean reprocessing of spent nuclear fuel, which is currently not permitted. The Korean Minister of the Knowledge Economy called for Korea to achieve "peaceful nuclear sovereignty" under future U.S.

agreements.[1] The current agreement expires in 2014, and the first discussions on renewal are likely within the next year. Congress will have an opportunity to review any new agreement before it takes effect.

South Korea's nuclear technology progression has been similar to the earlier nuclear paths of France and Japan, and which appears likely to be followed in the future by China. France, Japan, and now South Korea developed their nuclear power industries with technology and designs licensed from U.S. companies to supply domestic energy needs. In each case, the licensees assumed progressively greater responsibility for construction of the U.S.-designed units and eventually the engineering and design as well. The foreign firms now compete for nuclear plant contracts throughout the world, including the United States, either in consortia with their former U.S. licensors or independently.

In the UAE deal, the South Korean consortium is headed by government-owned Korea Electric Power Corporation (KEPCO) and includes other major Korean industrial companies that are involved in Korea's rapidly growing domestic nuclear power plant construction program. The consortium also includes Pittsburgh-based Westinghouse Electric Company, which currently owns the U.S. design on which the Korean design is based, and the Japanese industrial conglomerate Toshiba, which now owns most of Westinghouse.

Although Korean companies now take the lead on design and construction of Korea's nuclear power plants, Westinghouse still provides support under the design license. Such support typically includes components, instrumentation and control equipment, and technical and engineering services. The Korean plant to be built in the UAE, the APR-1400 model, is a modified version of the System 80+ design developed by the U.S. firm Combustion Engineering (C-E), which was later acquired by Westinghouse. Westinghouse has estimated that its work under the UAE contract could total $1 billion.[2]

DOMESTIC SOUTH KOREAN NUCLEAR ENERGY PROGRAM

When the construction of South Korea's first commercial nuclear power plant began in 1972, the South Korean economy was about 7.5% the size of Japan's, and the country's per-capita income was slightly lower than that of North Korea.[3] With such a relatively small industrial base, South Korea's plans to finance and operate a fleet of nuclear power plants could have been considered overly ambitious, and its long-term plans to master the new technology might have seemed unrealistic. But the subsequent growth of the South Korean economy—with per-capita income now rivaling other developed nations and GDP nearly 20% of Japan's—turned out to be more than sufficient to sustain the country's planned nuclear power development.

South Korea launched its nuclear power program through the government-owned Korea Electric Company (now Korea Electric Power Corporation, KEPCO), which purchased the country's first nuclear power units from Westinghouse. Those first plants were ordered on a turn-key basis, in which the foreign supplier delivered a completed plant with minimal Korean industry input. As shown in **Figure 1**, the first of these turn-key units, Kori 1, began operating in 1978. Westinghouse supplied the reactor and other components of the nuclear steam supply system (NSSS) and constructed the plant, and other western firms provided the turbine-generator and architect/engineering services. Wolsong 1 and Kori 2, coming on line in

1983, were also turn-key units, with all major components and construction services provided primarily by non-Korean companies.

Year of Operation/ Reactor Name	Reactor Supplier	Generator Supplier	Architect Engineer	Construction Contractor
1978 Kori 1	Westinghouse	GE (UK)	Gilbert	Westinghouse
1983 Wolsong 1	AECL	NE (UK)	AECL	AECL
1983 Kori 2	Westinghouse	GE (UK)	Gilbert	Westinghouse
1985 Kori 3	Westinghouse	GE (UK)	Bechtel	Hyundai
1986 Kori 4	Westinghouse	GE (UK)	Bechtel	Hyundai
1986 Yonggwang 1	Westinghouse	Westinghouse	Bechtel	Hyundai
1987 Yonggwang 2	Westinghouse	Westinghouse	Bechtel	Hyundai
1988 Ulchin 1	Framatome	Alstom	Framatome	Dong Ah/Hanjung
1989 Ulchin 2	Framatome	Alstom	Framatome	Dong Ah/Hanjung
1995 Yonggwang 3	Hanjung/C-E	Hanjung/GE	KOPEC/S&L	Hyundai
1996 Yonggwang 4	Hanjung/C-E	Hanjung/GE	KOPEC/S&L	Hyundai
1997 Wolsong 2	AECL/Hanjung	Hanjung/GE	AECL/KOPEC	Hyundai
1998 Wolsong 3	AECL/Hanjung	Hanjung/GE	AECL/KOPEC	Daewoo
1998 Ulchin 3	Hanjung/C-E	Hanjung/GE	KOPEC/S&L	Dong Ah/Hanjung
1999 Wolsong 4	AECL/Hanjung	Hanjung/GE	AECL/KOPEC	Daewoo
1999 Ulchin 4	Hanjung/C-E	Hanjung/GE	KOPEC/S&L	Dong Ah/Hanjung
2002 Yonggwang 5	Doosan	Doosan	KOPEC	Hyundai/Daelim
2002 Yonggwang 6	Doosan	Doosan	KOPEC	Hyundai/Daelim
2004 Ulchin 5	Doosan	Doosan	KOPEC	Dong Ah/Doosan/Samsung
2005 Ulchin 6	Doosan	Doosan	KOPEC	Dong Ah/Doosan/Samsung

Non-Korean Companies | Non-Korean and Korean Companies | Korean Companies

AECL: Atomic Energy of Canada Ltd. KOPEC: Korea Power Engineering Company
C-E: Combustian Engineering NE: Northern Engineering
GE: General Electric S&L: Sargent & Lundy

Source: Nuclear News, World List of Nuclear Power Plants, March 2009.

Figure 1. South Korean Nuclear Power Units.
Major Categories of Work Conducted By Domestic and Foreign Companies.

After those first three units, Korean firms took over the construction work on subsequent plants, beginning with Kori 3, which began commercial operation in 1985. However, the NSSS, turbine- generators, and architect/engineering services continued to be provided primarily by non-Korean companies, including Westinghouse, Atomic Energy of Canada Limited (AECL), and the French firm Framatome, which had previously licensed its design from Westinghouse. That arrangement continued for the next six units, which came on line from 1995-1999.

In 1987, KEPCO embarked on an effort to establish a standard Korean design, selecting the System 80 design from the U.S. firm Combustion Engineering (C-E) as the basis.[4] The System 80 design had been used for three identical units nearing completion at Palo Verde, Arizona.

Combustion Engineering won the competition for the Korean standard design contract by agreeing to full technology transfer, according to KEPCO. The U.S. contractors for the turbine- generators (General Electric) and architect/engineering services (Sargent & Lundy) agreed to transfer key technology as well.

All of the major U.S. companies working on the new C-E plants agreed to serve as subcontractors to Korean firms: Combustion Engineering to Hanjung (now Doosan), General Electric (GE) also to Hanjung, and Sargent & Lundy (S&L) to Korea Power Engineering Company (KOPEC), which is majority owned by KEPCO.[5] Yonggwang 3 and 4 and Ulchin 3 and 4, which began operating from 1995 to 1999, were constructed under that arrangement. A similar partnership was formed with Atomic Energy of Canada Limited (AECL) to build Wolsong 2-4, completed during 1997 through 1999. Plant construction continued to be carried out by Korean companies. Components and work shared by Korean and foreign firms is shown in **Figure 1**.

Korean reactor designers worked with C-E, which became part of Westinghouse in 2000, to develop a standard Korean design from the System 80 model. This effort resulted in the 1,000 megawatt Optimized Power Reactor (OPR-1000). The four OPR-1000 units that have been completed since 1999, Yonggwang 5-6 and Ulchin 5-6, were been built and constructed almost entirely by Korean firms. However, some key components continued to be supplied or heavily supported by non-Korean firms, constituting a small percentage of each nuclear unit's total cost.

Development of a larger and more advanced model of the Korean standard design was based on the C-E System 80+ design that received U.S. standard design certification from the Nuclear Regulatory Commission (NRC) in May 1997. The Korea Atomic Energy Research Institute (KAERI) helped develop the U.S. design,[6] complementing work on the Korean version of the design that began in 1992. The Korean design was largely completed by 1999 and was designated the APR-1400.[7]

South Korea currently has eight nuclear units under construction, four OPR-1000s and four APR- 1400s, to be completed between 2010 and 2016. Four more APR-1400s are to be completed between 2018 and 2021. That construction program would increase the country's nuclear power reactors from 20 to 32, and nuclear power generating capacity from 17,700 megawatts to 32,500 megawatts. South Korea's long-term electricity plan calls for increasing nuclear capacity to 42,700 megawatts by 2030, the equivalent of about seven additional APR-1400s after 2021.[8] South Korea generated 36% of its electricity from nuclear plants in 2008 and plans to increase that share to 59% by 2030.[9]

Ever since South Korea completed the first nuclear unit in which Korean firms participated in all phases of development—Yonggwang 3 in 1995—the country has opened an average of about one unit every 18 months. Starting after the most recent reactor began commercial operation in 2005, Ulchin 6, South Korea plans to complete an average of about one reactor every 16 months through 2021, and potentially about one reactor per year from 2022 to 2030. With the planned rate of domestic nuclear plant construction remaining fairly stable, it would appear that any significant expansion of South Korea's nuclear engineering and construction industry would depend on exports.

South Korean Nuclear Plant Export Program

The South Korean government expects KEPCO's reactor sale to the UAE to constitute the leading edge of a much larger nuclear power marketing effort throughout the world. According to news media reports, the Ministry of Knowledge Economy (MKE), which is responsible for industrial and trade policy, has established a goal for South Korea to capture 20% of the world nuclear power plant market during the next 20 years. Based on an estimated world market of about 400 large commercial reactors through 2030, a 20% penetration would result in South Korean exports of 80 reactors during the next 20 years, with an estimated value of $400 billion.[10]

UAE Reactor Contract

As South Korea's first foreign reactor sale, the UAE contract is likely to establish a template for future exports. The companies involved in the UAE project appear to be the same ones that are currently building Korean domestic nuclear plants and are likely to play similar roles in the export program. The importance placed by the ROK government on the contract was underscored by the presence of South Korean President Lee Myung-bak at the signing ceremony in the UAE December 27, 2009, along with UAE President Sheikh Khalifa bin Zayed al-Nahayan.[11]

The selection of the KEPCO consortium was made by the Emirates Nuclear Energy Corporation (ENEC), which will oversee the contract's implementation. According to a statement issued by ENEC, the contract includes the following major provisions:[12]

- The KEPCO consortium will design, build, help operate and maintain, and provide initial fuel for four APR-1400 nuclear units at a total cost of about $20 billion. A "high percentage of the contract" will be under a fixed price.
- Korean investors will have an equity interest in the UAE plants.
- The first unit is to begin commercial operation in 2017, with the other three to be completed by 2020.
- Extensive training, human resources development, and education is to be provided to allow UAE to eventually provide most of the nuclear plant staffing and develop commercial infrastructure and support businesses.
- A potential follow-on contract for long-term operation and maintenance of the units, worth as much as another $20 billion over 60 years,[13] is under discussion with KEPCO and other vendors.

The Korean consortium was selected over two other proposals, from Areva and General Electric- Hitachi. According to media reports, the decision was strongly affected by price. One report indicated that the KEPCO total of $20 billion was 30% lower than the Areva bid, which in turn was lower than the GE-Hitachi offer.[14] Another report described KEPCO's price as $16 billion lower than Areva's.[15] KEPCO's bid averages out to $5 billion per reactor, which is higher than a reported estimate of $3.15 billion for each of two APR-1400s being built at the Shin-Kori site in Korea, with the difference ascribed to the additional costs of operating in a country with no previous nuclear experience.[16] The $20 billion cost of 4,200 megawatts of electric generating capacity works out to $3,571 per kilowatt, which is similar to the average of recent comparable U.S. estimates for new nuclear plants. (See CRS Report RL34746, *Power Plants: Characteristics and Costs*, by Stan Mark Kaplan.)

Korean and U.S. Partnership

As noted above, the Korean and U.S. companies involved in the UAE project have worked together extensively in the past on the domestic Korean nuclear power program, with Korean firms gradually taking over most of the work. Below are the members of the KEPCO consortium and their roles in the UAE project:

- *KEPCO*. Prime contractor and project integration. According to the ENEC announcement, "KEPCO is expected to begin mobilizing its team in Abu Dhabi immediately."
- *Korea Hydro and Nuclear Power Company (KHNP)*. Operating company for Korean nuclear power plants. To serve as engineering, procurement, and construction contractor and operator. KEPCO subsidiary.
- *KOPEC*. Nuclear power plant architect/engineering services. Majority owned by KEPCO.
- *Korea Nuclear Fuel Company (KNF)*. Initial nuclear fuel loads.
- *Korea Plant Service and Engineering Company (KPS)*. Plant maintenance.[17] Majority owned by KEPCO.
- *Doosan Heavy Industries & Construction*. Fabrication of nuclear steam supply system and other major components.
- *Samsung C&T Corporation*. Plant construction.
- *Hyundai Engineering and Construction*. Plant construction.
- *Westinghouse Electric Company*. Technical and engineering support services and various components.
- *Toshiba Corporation*. Majority owner of Westinghouse. Role unspecified. Possible component supply and technical consulting.

Westinghouse has projected that its share of the contract will be as much as $1 billion, or about 5% of the total, generating or sustaining several hundred U.S. jobs. Westinghouse has not released the specific scope of its activities under the contract, except that it may include the supply of plant components, man-machine interface systems, instrumentation and control equipment, and technical and engineering support services.[18] The estimated Westinghouse share of the UAE contract is similar to its reported share of the construction of the two Shin-Kori APR- 1400s, about $300 million. Specific reactor components being provided by Westinghouse under the 2006 Shin-Kori contract include reactor coolant pumps and motors, reactor vessel internals, and control element drive mechanisms.[19]

Although most of the U.S. technology involved in the Korean standard reactor designs (OPR-1000 and APR-1400) has been successfully transferred to Korean firms as called for by the initial C-E licensing agreement in 1987, Westinghouse still considers the Korean reactors to be Westinghouse-licensed products. As a result, the UAE reactor sale will be subject to U.S. export control requirements.[20]

Westinghouse is also partnering with Korean industry to produce control element assemblies for C-E reactor designs, including U.S. and Korean nuclear power plants, as well as the UAE plants and other potential Korean exports. The joint venture between Westinghouse and KNF is being located at the KNF fuel fabrication plant in Daejeon, Korea, and will bc 55% owned by Westinghouse.[21]

Outlook for Korean Exports

The goal set by the Korean Ministry of Knowledge Economy for a 20% South Korean share of the global nuclear power plant market would place South Korea about equal to Russia and behind only France and the United States in the nuclear market, according to a ministry report to President Lee. "Nuclear power-related business will be the most profitable market after automobiles, semiconductors, and shipbuilding," the report said. MKE also called for South Korean firms to expand their share of the estimated $78 billion world market for operation, maintenance, and repair of nuclear power plants. MKE announced an interim goal of signing contracts to export six more Korean nuclear units, in addition to the four purchased by the UAE, by 2012.[22]

The UAE contract added substantial credibility to MKE's export goals and changed the dynamics of the world nuclear power market. Kuwaiti officials were paraphrased as saying the UAE price "is likely to become a benchmark for atomic energy technology across the region."[23] After losing the UAE contract, Areva was reported to be examining ways to modify its plant design to cut costs, such as by cutting the number of steam generators from four to two by making them larger, as in the Korean design.[24]

Other potential Korean export deals are under consideration in Indonesia and the United States, Westinghouse's home market. Other countries that have been mentioned in the news media include Vietnam, Malaysia, Thailand, and Middle East neighbors of the UAE.[25] A Korean consortium has reportedly been selected by Jordan to build the kingdom's first research reactor.[26] KNHP signed a preliminary agreement with Indonesian energy firm Medco Energi in July 2007 to build Korean-design reactors in Indonesia.[27] A feasibility study is currently underway for four OPR-1000 units that could begin operating by 2016.[28] However, Japanese firms are reportedly also under consideration.[29] A U.S. energy development company, Alternate Energy Holdings Incorporated (AEHI), announced in January 2010 that it was negotiating with South Korean officials on an agreement to build APR-1400 nuclear units at proposed sites in Idaho and Colorado. An AEHI news release said, "We expect the agreement to be similar to the UAE agreement announced last week. Such technology should give AEHI a serious competitive advantage."[30] Plans for financing for the proposed projects are unknown, however.

KEPCO took an early step toward exporting the APR-1400 to the United States by meeting with the U.S. NRC November 18, 2009, on possible standard design certification for the reactor. At the "initial pre-application meeting," KEPCP gave a presentation on the Korean nuclear industry, the U.S. and Korean work on developing the System 80+ and the APR-1400, and differences between the two designs. Prospects for NRC certification of the APR-1400 could presumably be helped by its similarity to the previously certified System 80+. However, NRC officials stressed that "this meeting did not initiate the review of the APR-1400 design certification and that staff resources for this review are not currently in the NRC budget for at least the next two years. Additional interactions with KEPCO regarding their plans are expected to occur in the future."[31]

The capacity of the Korean nuclear industry would apparently need to expand to meet MKE's export goals. As noted above, Korea has been completing an average of about one reactor every 16 months and plans to increase that rate to about one per year after 2021 for the domestic market. In addition, Doosan and other firms have been producing major reactor components for non-Korean reactors, such as the four Westinghouse AP-1000 units being built in China. To export 80 units by 2030, as implied by the MKE goal, the Korean industry would have to complete an additional four units per year, a substantial increase over the current rate. However, the total implied construction rate of about five units per year has been achieved by other countries in the past, such as France during the 1980s.[32]

To expand Korea's nuclear plant construction and service capacity, MKE has announced plans to train 2,800 nuclear technical staff by 2011 and invest $350 million in further design improvements, including an increase in research and development personnel. Under the MKE plan, South Korea would become completely self-sufficient in nuclear reactor technology by 2012.[33]

U.S. Policy Considerations

The future direction of U.S.-South Korean cooperation in world energy markets poses a number of near- and long-term policy considerations for the United States. U.S. policymakers will face decisions related to U.S. nuclear energy cooperation with Korea that will affect broader U.S. policy goals. In turn, U.S. decisions based on broad policy goals will have an effect on Korean involvement with the U.S. nuclear industry.

The most immediate potential decision related to these issues will be the Secretary of Energy's judgment on whether to authorize Westinghouse and its employees and contractors to participate in the KEPCO reactor sale to the UAE. The so-called 810 Authorization, under Title 10, Part 810, of the U.S. Code of Federal Regulations, is required for any U.S. person to assist in the production of special nuclear materials (such as through construction and operation of a nuclear power plant) in another country. According to Westinghouse officials, "Westinghouse will submit the required application to the U.S. DOE for technology export authorization and Westinghouse will be involved in the project as the originator of the technology."[34]

Section 810 implements Section 57b of the Atomic Energy Act (42 U.S.C. 2077), which also prohibits nuclear technology export to countries without an agreement for nuclear cooperation under Atomic Energy Act Section 123. Under Section 57b, the Secretary of Energy must determine that the proposed technology export "will not be inimical to the interest of the United States" with the concurrence of the Department of State and after consultation with NRC and the Departments of Commerce and Defense. The recently implemented U.S.-UAE 123 agreement excludes fuel cycle facilities, such as uranium enrichment and reprocessing plants, that could produce weapon-useable material. South Korea signed its own nuclear cooperation agreement with the UAE on June 22, 2009. (For more information about the U.S.-UAE 123 agreement, see CRS Report R40344, *The United Arab Emirates Nuclear Program and Proposed U.S. Nuclear Cooperation*, by Christopher M. Blanchard and Paul K. Kerr.)

Because of the restrictions in the U.S.-UAE 123 agreement, the 810 authorization for the South Korean reactors is likely to raise fewer issues than it might have otherwise. Nevertheless, the authorization will undergo a thorough interagency review relating to U.S. interests in the region and throughout the world, such as controls on the re-transfer of U.S.-origin nuclear technology after it is exported by Korea. Future South Korean exports will be subject to similar 810 reviews as long as U.S. companies or individuals are involved.

The degree to which the United States retains control over the South Korean nuclear program has led to proposals for South Korean "nuclear sovereignty." As reported in the Korean news media, the term has been applied to restrictions ranging from the current U.S. prohibition on South Korean spent fuel reprocessing to approvals of technology exports, such as the UAE sale. South Korean Minister of Knowledge Economy Choi Kyung-hwan recently clarified that the term does not refer to nuclear weapons development by using the phrase "peaceful nuclear sovereignty." Choi called the current U.S. restrictions on Korean spent fuel reprocessing "excessive," and pointed to the UAE sale as evidence of, as paraphrased in a Korean news report, "global confidence in South Korea's ability to handle the task."[35] KAERI is conducting a laboratory-scale research program on reprocessing spent fuel with an advanced pyroprocessing technique. However, the level of consensus over the pyroprocessing option among government agencies, Korean electric utilities, and the public remains uncertain.

The current U.S.-Korea 123 agreement prohibits South Korea from reprocessing any U.S.-origin spent fuel, including spent fuel from South Korea's U.S.-designed reactors, without U.S. permission. KAERI's pyroprocessing technology would partially separate plutonium and uranium from spent fuel, but the United States has not allowed the technology to be applied to actual spent fuel. The 123 agreement expires in 2014, and some South Korean officials have called for significant changes in the reprocessing restrictions, such as the granting of the advance approval already given to Japan and Western Europe. According to one Korean news account, the Korean reactor sale to the UAE "appears to have boosted efforts to seek a revision" of the U.S.-Korea 123 agreement.[36] Although the issue has the potential to become an irritant in bilateral relations between Washington and Seoul, thus far it appears to be a relatively minor one. As of early 2010, many experts contacted by CRS describe U.S.-South Korean relations as in their best shape in over a decade.

The United States and South Korea continue to cooperate on nuclear technology development in key international forums. Through the Global Nuclear Energy Partnership (GNEP), the United States, South Korea, and 23 other partner nations (plus 31 observers) are studying world needs for nuclear infrastructure development and reliable nuclear fuel supply, including reprocessing technology. The United States and South Korea are also founding members of the Generation IV International Forum (GIF), a cooperative international research effort on advanced nuclear reactor technology. Nuclear energy R&D was a high priority of the George W. Bush Administration. Under President Barack Obama, funding has continued for those efforts, although the programs have been refocused away from deployment and toward fundamental research. Recent U.S. news media reports have indicated that there may be disagreement within the Obama Administration over future funding for DOE nuclear energy R&D.[37]

Future growth of South Korea's nuclear energy technology export program would offer opportunities for the U.S. nuclear industry as well as significant challenges. As the KEPCO-UAE sale makes clear, U.S. companies can directly benefit from participation in Korean export projects, in this case an estimated $1 billion for Westinghouse. But the UAE project also illustrates the potential competition the U.S. nuclear suppliers may face from Korea, which overcame a GE-Hitachi bid in the final round and an earlier Westinghouse proposal. As the South Korean nuclear industry develops more reactor components of indigenous design, the opportunities for U.S. participation in South Korean export projects may diminish. Government ownership of KEPCO may also be a competitive concern for U.S. industry. A World Nuclear Association report notes that South Korea may develop a large, exportable reactor design based on the APR- 1400 with indigenous components by 2015, "though Westinghouse is not likely to let it compete in main markets such as USA and China without KEPCO buying the rights to the design."[38] Such issues related to South Korean "peaceful nuclear sovereignty" are likely to be a topic of continuing U.S.-ROK discussion.

End Notes

[1] "Seoul Wants 'Sovereignty' in Peaceful Nuclear Development," *Chosun Ilbo*, December 31, 2009.

[2] E-mail from Jeanne T. Lopatto, Vice President, Government and International Affairs, Westinghouse Electric Company, January 5, 2010.

[3] Angus Maddison, *The World Economy: Historical Statistics*, Organization for Economic Cooperation and Development, 2006, pp. 298-306.

[4] World Nuclear Association, South Korea Country Report, January 7, 2010, http://www.world-nuclear.org/info/inf8 1 .html.

[5] Gary Baker and Shin Ho-Chul, "Korean Utility and Three U.S. Firms Ink Nuclear Building Deal," *Nucleonics Week*, April 16, 1987, p. 5.

[6] National Academy of Sciences, *Nuclear Power: Technical and Institutional Options for the Future*, Washington, DC, 1992, p. 100.

[7] World Nuclear Association, op. cit.

[8] Korea Atomic Energy Research Institute, *Nuclear Power and R&D Programs in Korea*, Presentation to CRS visitors, Daejeon, South Korea, July 27, 2009, pp. 11-12.

[9] Jae-min Ahn, "Industry Perspectives on Korean Nuclear Program," Presentation to U. S.-ROK Workshop on Nuclear Energy and Nonproliferation, Washington, DC, January 20, 2010.

[10] World Nuclear News. "South Korea Weeks to Boost Reactor Exports," January 13, 2010, http://www.world-nuclearnews.org/NP-South_Korea

[11] Amena Bakr, "South Korean Group Wins $40bn UAE Nuclear Deal," *Arabian Business.com*, December 27, 2009.

[12] Emirates Nuclear Energy Corporation, "UAE Selects Korea Electric Power Corp. Team as Prime Contractor for Peaceful Nuclear Power Program," press release, December 27, 2009, http://www.enec.gov.ae/news/uae-selects-koreaelectric-power-corp-team-as-prime-contractor-f/.

[13] Bakr, op. cit.

[14] Chris Stanton, "Nuclear Bid to Be Industry Norm," *The National*, December 28, 2009, http://www.thenational.ae/apps/pbcs.dll/article?AID=/20091228/BUSINESS/712289928.
[15] Bakr, op. cit.
[16] Stanton, op. cit.
[17] "KEPCO to Provide Design, Construction and Maintenance of Nuclear Reactors," *Khaleej Times Online*, December 28, 2009.
[18] Lopatto, op. cit.
[19] "Two New Nuclear Plants for South Korea," *Nuclear Engineering International*, September 6, 2006.
[20] Lopatto, op. cit.
[21] April Murelio, "Westinghouse, KNF Team to Manufacture Control Element Assemblies," *Nuclear Power Industry News*, February 9, 2009.
[22] World Nuclear News, op. cit.
[23] Stanton, op. cit.
[24] Ann MacLachlan, "Areva Considering Ways of Cutting Costs of EPR Nuclear Reactor," *Platts Commodity News*, January 11, 2010.
[25] Global Collaborative, "KEPCO/KHNP," http://www.globalcollab.org/Nautilus/australia/reframing/aust-ind-nuclear/ ind-np/muria/countries/kepco.
[26] Taylor Luck, "S. Korean Bidder to Build Nuclear Research Reactor in Jordan," *Jordan Times*, April 12, 2009.
[27] "S. Korea, Indonesia Sign Nuclear Cooperation Deal," *Reuters*, July 25, 2007.
[28] World Nuclear Association, op. cit.
[29] Reuters, op. cit.
[30] Alternate Energy Holdings Incorporated, "AEHI Expects to Close Deal to Import Korean Reactors in Early 2010," press release, January 4, 2010, http://www.alternateenergyholdings.com/LinkClick.aspx?fileticket=gVVisrNAzgI%3d&tabid=1979.
[31] Nuclear Regulatory Commission, *Korea Electric Power Corporation APR-1 400 Presentation*, ML093430109, December 9, 2009, http://adamswebsearch.nrc.gov/scripts
[32] Arnulf Grubler, *An Assessment of the Costs of the French Nuclear PWR Program 1970-2000*, International Institute for Applied Systems Analysis, IR-09-036, Laxenburg, Austria, October 6, 2009, p. 8, http://www.docstoc.com/docs/ 19721910/Interim-Report-IR-09-036-An-assessment-of-th.
[33] World Nuclear News, op. cit.
[34] Lopatta, op. cit.
[35] "Seoul Wants 'Sovereignty' in Peaceful Nuclear Development," *Chosun Ilbo*, December 31, 2009.
[36] Chosun Ilbo, op. cit.
[37] George Lobsenz, "White House Moves to Restrict DOE Nuclear Research," *Energy Daily*, January 15, 2010, p. 1.
[38] World Nuclear Association, op. cit.

In: U.S. Nuclear Cooperation with Other Countries
Editor: Gabriel A. Oliveira
ISBN: 978-1-61209-910-1

Chapter 7

THE UNITED ARAB EMIRATES NUCLEAR PROGRAM AND PROPOSED U.S. NUCLEAR COOPERATION

Christopher M. Blanchard and Paul K. Kerr

SUMMARY

The United Arab Emirates (UAE) has embarked on a program to build civilian nuclear power plants and is seeking cooperation and technical assistance from the United States and others. During 2008 and early 2009, the Bush Administration and the UAE government negotiated and signed a memorandum of understanding and a proposed bilateral agreement on peaceful nuclear cooperation pursuant to Section 123 of the Atomic Energy Act (AEA) of 1954. Then-U.S. Secretary of State Condoleezza Rice signed the proposed agreement on peaceful nuclear cooperation with the UAE January 15, 2009. Deputy Secretary of State James Steinberg signed a new version of the agreement May 21; the Obama Administration submitted the proposed agreement to Congress the same day.

Under the AEA, Congress has the opportunity to review such a proposed agreement for 90 days of continuous session, after which the agreement becomes effective unless, during that time, Congress adopts a joint resolution disapproving the agreement and the resolution becomes law. According to the Office of the Parliamentarian of the House of Representatives, the 90 days of continuous session for the proposed U.S.-UAE agreement expired October 17, 2009. The UAE cabinet approved the agreement October 26. The agreement entered into force after the two governments exchanged diplomatic notes December 17, 2009. The agreement text states the intent of both governments to cooperate in a number of areas including, but not limited to, the development of the UAE's "civilian nuclear energy use in a manner that contributes to global efforts to prevent nuclear proliferation" and, "the establishment of reliable sources of nuclear fuel for future civilian light water reactors deployed" in the UAE.

Some Members of Congress had welcomed the UAE government's stated commitments not to pursue proliferation-sensitive nuclear capabilities, such as uranium enrichment or spent fuel reprocessing. Other Members had signaled their intention to weigh the proposed bilateral agreement in light of parallel and specific concerns about the UAE's cooperation with international efforts (such as sanctions) to prevent Iran from developing nuclear weapons and ballistic missiles, as well as the potential proliferation or safety risks inherent to exporting U.S. nuclear technology.

In the 111th Congress, legislation (H.R. 364) has been introduced that would have required President Obama to certify that the UAE had taken a number of steps to strengthen its export controls and stem illicit trade with Iran before any agreement could come into effect or related U.S. exports of nuclear technology to the UAE could be approved. In 2007, the UAE adopted a stronger export control law, but has yet to issue implementing regulations for the law or to fully staff a national export control body to enforce it. In the interim, export control enforcement functions remain the responsibility of authorities in the UAE's individual emirates, in coordination with a new national interagency Committee on Commodities Subject to Import and Export Control established in April 2009. According to UAE officials, cooperation with the United States has resulted in a number of joint interdiction operations. This report provides background information on the UAE nuclear program, reviews developments to date, analyzes proposed nuclear cooperation with the United States, and discusses relevant legislative proposals and options. See also CRS Report RS21852, *The United Arab Emirates (UAE): Issues for U.S. Policy*, by Kenneth Katzman, and CRS Report RS22937, *Nuclear Cooperation with Other Countries: A Primer*, by Paul K. Kerr and Mary Beth Nikitin.

THE UNITED ARAB EMIRATES: BACKGROUND AND U.S. RELATIONS

The United Arab Emirates (UAE) is a federation of seven emirates (principalities): Abu Dhabi, Dubai, Sharjah, Ajman, Fujayrah, Umm Al Qawayn, and Ras Al Khaymah. National authority rests in the hands of a Federal Supreme Council, which is composed of the hereditary rulers of the country's constituent emirates and elects the national president from among its members. Sheikh Khalifa bin Zayed Al Nahyan, the ruler of Abu Dhabi, was elected UAE President in 2004 following the death of his father Sheikh Zayed bin Sultan Al Nahyan, who had ruled Abu Dhabi since 1966 and served as UAE President since 1971. In practice, the wealthier and more powerful emirates of Abu Dhabi and Dubai exercise the strongest influence over the country's affairs; under current convention, the ruler of oil-rich Abu Dhabi serves as the UAE President, and the ruler of the UAE's commercial hub, Dubai, serves as Vice President. The Supreme Council appoints the Prime Minister and the Council of Ministers (cabinet), which initiates legislation for ratification by the Supreme Council and the President.

The United States and the UAE have enjoyed close and cooperative relations in recent years, in spite of periodic differences with regard to political reform, the Israel-Palestinian conflict, counterterrorism, and U.S. policies regarding Iraq and Iran. Military cooperation and arms sales form a key pillar of U.S.-UAE relations. The UAE hosts frequent port calls and shore visits for U.S. naval vessels and allows the U.S. military to use Al Dhafra air base in support of a variety of missions in the U.S. Central Command (CENTCOM) area of operations. In 2007 and 2008, the Bush Administration notified Congress of over $19.4

billion in potential arms sales to the UAE, including what would be the first overseas sale of the Terminal High Altitude Air Defense system.

Bilateral trade has increased in recent years, with 2008 U.S. exports valued at over $15.7 billion, making the UAE the largest U.S. export market in the Middle East. The Bush Administration began negotiating a free trade agreement with the UAE in 2004, but did not conclude the negotiations. The United States does not import a significant amount of oil from the UAE. However, the UAE exports over 2 million barrels of oil per day, making it a key global energy producer.

THE UNITED ARAB EMIRATES NUCLEAR PROGRAM

The government of the United Arab Emirates (UAE), like others in the Middle East, has announced plans to acquire nuclear energy production technology as a means of meeting projected national energy consumption needs.[1] Renewed global interest in nuclear power has led some experts and observers to express concern that the projected spread of nuclear technology in coming years could contribute to nuclear proliferation. In the Middle East, added scrutiny is often applied to the motives and choices of regional actors regarding nuclear technology because of concern that Iran's nuclear program and Israel's presumed nuclear weapons may motivate other regional governments to seek nuclear technology for strategic or military purposes. Other concerns about nuclear safety relate to potential terrorist attacks or political instability, both of which have threatened some regional countries in recent years. UAE officials report that they have considered these potential risks carefully, and have announced plans and measures intended to address proliferation and security concerns.

While UAE officials stress that final decisions about the scope of the nuclear program and potential international participation have not been made, plans for the program are at an advanced stage and are awaiting final legal authorization. Policymakers and advisers in the government of Abu Dhabi, in consultation with representatives from the other six emirates, have set out an ambitious agenda for the program and are guiding its preliminary implementation. In April 2008, the UAE government issued a policy statement[2] that provides a rationale for the country's perceived need for nuclear energy and states guiding principles for the nuclear energy program. Nascent operating and regulatory bodies have been formed and are awaiting the passage of authorizing legislation to begin their formal work. The end goal of the program, according to officials and related documents, is to build and operate a "fleet" of nuclear power plants to generate electricity for the UAE, supported by advanced, indigenously managed safety, regulatory, and security agencies.

The UAE government pledged $10 million in August 2008 toward an international nuclear fuel bank proposed by the Nuclear Threat Initiative, a non-governmental organization. The bank would be administered by the International Atomic Energy Agency (IAEA),[3] which would also implement safeguards on any future UAE nuclear facilities. Mr. Hamad Al Kaabi, UAE Special Representative for International Nuclear Cooperation, explained August 7 that the contribution is part of the UAE's policy to support multilateral fuel supply efforts—a policy consistent with the country's decision to rely on foreign fuel suppliers.

Rationale

UAE officials estimate that their country must expand its power generation and transmission capacity from the current level of 16 gigawatts to 40 gigawatts by 2020 in order to meet projected demand increases, which they estimate will continue growing at a 9% annual rate. In spite of the recent slowdown in global and domestic economic activity, representatives of the UAE nuclear program believe that the energy demand projections they are using to justify and plan the acquisition of nuclear plants remain accurate, particularly in light of planned industrial and commercial projects in energy-intensive sectors in the emirate of Abu Dhabi.[4] To date, UAE officials and representatives have not publicly shared economic cost and energy use analyses referred to in briefings on their nuclear program.

In arguing for nuclear energy as a solution to the country's projected energy needs, the UAE government policy statement concludes that "known volumes of natural gas that could be made available to the nation's electricity sector would be insufficient to meet future demand."[5] The UAE currently exports roughly 600 million standard cubic feet per day of natural gas to Japan under long-term supply arrangements and imports roughly 2 billion cubic feet of natural gas from Qatar via the underwater Dolphin pipeline system.[6] Similarly, UAE officials believe that crude oil and diesel could be "logistically viable" sources of energy, but would impose high economic opportunity costs (as a result of lost export revenue) and environmental costs. Officials determined that coal could be a more economical solution, but would have even greater environmental costs and, as an import, also would raise concerns for the UAE about the security of supply. UAE officials believe that solar and wind energy sources could supply "only 6-7% of peak electricity demand by 2020," even after "aggressive development."

Development Plans

The end goal of the program, according to officials and related documents, is to build and operate a "fleet"[7] of nuclear power plants to generate electricity for the UAE, supported by advanced, indigenously managed safety, regulatory, and security agencies that will be developed over time and with outside assistance. The UAE government is seeking to bring its first nuclear power plant online by 2017 along with required facilities and equipment for safety, storage, and system management. Under current plans, capacity would expand thereafter to include multiple nuclear power plants. Reports suggest that the UAE is moving forward with a contract bidding and award process, with the goal of selecting a primary contractor or team in late 2009. UAE officials reportedly plan to build in contractor incentives for on-time delivery of a turnkey plant before the ambitious 2017 deadline.[8] No specific decisions have been made regarding the source of nuclear fuel for the planned nuclear reactor or on handling spent reactor fuel.

Current Infrastructure and Regulatory Regime

The UAE currently has no nuclear material under IAEA safeguards. It signed the NPT in 1995 and completed a Small Quantities Protocol in 2003, which applies to non-nuclear weapons states that do not have significant nuclear programs or nuclear material. The UAE also has undertaken Technical Cooperation projects with the Agency, some of which are directly related to nuclear electricity generation. For example, a project begun in 1977 advised the government "on the establishment of a nuclear energy administration." A 1984 project focused on uranium exploration. More recently, a Technical Cooperation project approved in 2005 was designed to assess the "technical and economic feasibility" of a nuclear power and desalination plant. Active IAEA Technical Cooperation projects with the UAE focus on human resources development for atomic energy, feasibility studies for waste management, environmental monitoring, and nuclear accident early warning preparedness and response.[9]

A national law authorizing the program was adopted by the Federal Supreme Council in early October 2009. According to the State Department, the law, Federal Law 6 of 2009, "prohibits uranium enrichment and spent fuel reprocessing, creates a Federal Authority for Nuclear Regulation (FANR), and develops a nuclear material licensing and control system."[10] UAE government representatives report that the UAE sought and received input on its draft nuclear law from the United States, United Kingdom, Japan, Korea, and France.[11] In conjunction with the issuance of the law, a Board of Management for the FANR was announced, and the board named former U.S. Nuclear Regulatory Commission Executive Director for Operations Dr. William Travers as the FANR's first Director General. A nuclear energy policy advisory board reportedly has been formed, and UAE officials report that its members, though unnamed, include leading international nuclear energy industry officials.[12] The UAE has also adopted a law governing export controls, but has not yet issued implementing regulations (see section on "Export Control Concerns"). The UAE has also stated that it intends to establish a "separate nuclear liability regime for third-party compensation modeled on the four as-yet un-ratified IAEA instruments on nuclear liability."[13]

Some limited consulting and contracting between U.S. firms and the UAE related to the UAE's proposed nuclear program has already taken place. In August 2008, Virginia's Thorium Power Ltd. signed two consulting and advisory services contracts related to the establishment of the Abu Dhabi-based Emirates Nuclear Energy Corporation (ENEC)[14] and the FANR. In October 2008, ENEC announced that Colorado's CH2M Hill, Inc. was selected for a 10-year contract as the managing agent for the evaluation and design stage of the nuclear energy program. Pennsylvania-based Rizzo and Associates Inc., has been hired to survey potential nuclear plant sites in the UAE. The contracts were signed with the government of Abu Dhabi.

Proposed U.S.-UAE Cooperation

Emirates Nuclear Energy Corporation (ENEC) officials have expressed interest in establishing joint ventures with U.S. firms for the design, construction, and operation of nuclear power plants. ENEC also is seeking to sign contracts for the specialized training of operations, security, and regulatory personnel in order to expand the indigenous human capital base within the UAE for independent management of planned nuclear facilities. During 2008 and early 2009, the Bush Administration and the UAE government negotiated and signed a Memorandum of Understanding (MOU) (see below) and a proposed bilateral agreement on peaceful nuclear cooperation pursuant to Section 123 of the Atomic Energy Act (AEA) of 1954. The nuclear cooperation agreement entered into force after the two governments exchanged the relevant diplomatic notes December 17, 2009.

Memorandum of Understanding

On April 21, 2008, the United States and the UAE signed a MOU "Concerning Cooperation in Peaceful Uses of Nuclear Energy." The MOU states that the two countries "intend to cooperate, subject to their respective national laws," in a variety of nuclear activities. The MOU is a statement of intent regarding future cooperation, but is not legally binding. Although such memoranda are not prerequisites for concluding future nuclear cooperation agreements, the State Department has argued that they are useful tools for cooperating with countries which are interested in the responsible use of nuclear energy because they create opportunities to solicit specific commitments with regard to safeguards and technology choices. An April 21, 2008, State Department press release described the U.S.-UAE MOU as a "tangible expression of the United States' desire to cooperate with states in the Middle East, and elsewhere, that want to develop peaceful nuclear power in a manner consistent with the highest standards of safety, security and nonproliferation." A similar MOU was concluded with Saudi Arabia in May 2008.

Proposed Bilateral Agreement Pursuant to Section 123 of the Atomic Energy Act of 1954

On May 21, 2009, Deputy Secretary of State James Steinberg and UAE Ambassador to the United States Yousef Al Otaiba signed the text of a bilateral agreement on peaceful nuclear cooperation. Although then-Secretary of State Condoleezza Rice and UAE Foreign Minister Abdullah bin Zayed Al Nahyan signed a similar agreement this past January, the two governments reopened the text for negotiation after the Obama administration took office.

Under the Atomic Energy Act of 1954 (AEA), all significant nuclear cooperation with other countries requires a peaceful nuclear cooperation agreement. Such agreements, which require congressional approval, are "framework" agreements which set the terms of reference and provide authorization for cooperation. The AEA includes requirements for an agreement's content, presidential determinations, and other supporting information to be submitted to Congress, conditions affecting the implementation of an agreement once it takes effect, as well as procedures for Congress to consider and approve the agreement (see "Nuclear Cooperation Agreements, Approval Process, and Proposed Changes" below). The agreement would enter into force on the date when the two governments "exchange diplomatic notes informing each other that they have completed all applicable requirements." As noted, the two governments exchanged these notes December 17, 2009.

According to the proposed U.S.-UAE agreement, the two countries "intend to cooperate" on a variety of nuclear activities, including

- Developing "requirements for grid-appropriate power reactors and fuel service arrangements;"
- Promoting the "establishment of a reliable source of nuclear fuel for future civil light water nuclear reactors;"
- "Civil nuclear energy training, human resource and infrastructure development;"
- Cooperating on nuclear security and nonproliferation, "including physical protection, export control and border security;"
- Developing the UAE's "civil nuclear energy use in a manner that supports global efforts to prevent nuclear proliferation, including, for example, the Global Nuclear Energy Partnership;"[15]
- Applying "radioisotopes and radiation in industry, agriculture, medicine and the environment;"
- Managing "radioactive waste and spent fuel;" and
- Identifying "uranium mining and milling resources."

According to the agreement, cooperation could include

- "Exchange of scientific and technical information and documentation;"
- "Exchange and training of personnel;"
- "Organization of symposia and seminars;"
- "Provision of relevant technical assistance and services;"
- Transfers of "material, equipment and components."

The agreement contains a variety of provisions which are required by the AEA and are designed to ensure that the UAE's nuclear program remains exclusively for peaceful purposes. It also includes two provisions which are not found in any other U.S. nuclear cooperation agreement. First, the agreement provides that the UAE bring into force its Additional Protocol to its IAEA safeguards agreement before the United States licenses "exports of nuclear material, equipment, components, or technology" pursuant to the agreement. The IAEA Board of Governors approved the Protocol March 3. The UAE signed it April 8, but has not yet brought it into force. Such protocols give IAEA officials greater access to an NPT state's nuclear-related facilities and information.

Second, the agreement states that the UAE

> shall not possess sensitive nuclear facilities within its territory or otherwise engage in activities within its territory for, or relating to, the enrichment or reprocessing of material, or for the alternation in form or content (except by irradiation or further irradiation or, if agreed by the Parties, post-irradiation examination) of plutonium, uranium 233, high enriched uranium, or irradiated source or special fissionable material.

A May 21 letter to Congress, which President Obama submitted along with the agreement, described this provision as a "legally binding obligation." According to the Nuclear Proliferation Assessment Statement submitted with the agreement, this provision "survives any termination of the Agreement so long as nuclear items subject to the Agreement remain in the territory of the UAE or under its jurisdiction or control anywhere." Furthermore, the agreement provides the United States with the right to terminate nuclear cooperation and to require the return of any nuclear "material, equipment or components ... and any special fissionable material produced through their use" if, after the agreement's entry into force, the UAE "possesses sensitive nuclear facilities within its territory or otherwise engages in activities within its territory relating to enrichment of uranium or reprocessing of nuclear fuel."[16]

Another provision, which is not typically included in nuclear cooperation agreements, requires both parties to give "due consideration ... to non-proliferation and physical protection aspects" when selecting a storage facility for special fissionable material.[17]

According to the agreement, the United States may also require that any special fissionable material that has been transferred to the UAE or "used in or produced through the use of any material or equipment" transferred pursuant to the agreement be transferred to either the United States or an unspecified "third country" if Washington "considers that exceptional circumstances of concern from a nonproliferation standpoint so require." A 1981 U.S. nuclear cooperation agreement with Egypt contains a similar restriction.

It is worth noting that an Agreed Minute to the U.S.-UAE agreement includes a provision which establishes its conditions as minimum standards for future such U.S. agreements in the Middle East. Stating that "the fields of cooperation, terms and conditions" accorded by the U.S.-UAE agreement "shall be no less favorable in scope and effect than those which may be accorded, from time to time, to any other non-nuclear-weapon State in the Middle East in a peaceful nuclear cooperation agreement," the Minute explains that, in the event that Washington concludes a more-favorable agreement with another regional government, the United States will, at the UAE's request, consult with the UAE "regarding the possibility of amending" the agreement in order to make its terms equally favorable to the new agreement.

A similar provision in the U.S.-Egypt agreement meant that the United States had to ensure that the agreement with the UAE would be at least as stringent. Since the latter agreement is more stringent, it has established a higher standard for future such U.S. agreements in the region.[18] (See the **Appendix**).

Additionally, the U.S.-UAE agreement provides a potential way for the UAE to transfer spent nuclear fuel to other countries. The Agreed Minute states that the UAE may transfer spent nuclear fuel to France or the United Kingdom for storage or reprocessing. In the past, such advance U.S. consent has been given only to Japan, Switzerland, and Norway.[19] The transferred material is to be held within EURATOM, and any separated plutonium cannot be returned to the UAE without additional U.S. consent.[20] [21] According to the agreement, approval for such UAE spent fuel transfers would be subject to several conditions, including the UAE's adherence to its declared policy of refraining from enrichment and reprocessing. The UAE may also not engage in fabricating nuclear fuel containing plutonium. Additionally, the United States can terminate an agreement regarding spent fuel transfers if Washington decides that the UAE has not met one of the relevant conditions or if the United States "considers that exceptional circumstances of concern from a non-proliferation or security standpoint so require." The agreement explains that "[s]uch circumstances include, but are not limited to, a determination ... that the approval cannot be continued without a significant increase of the risk of proliferation or without jeopardizing its national security."

Nuclear Cooperation Agreements, Approval Process, and Proposed Changes

As noted, all significant nuclear cooperation[22] with other countries requires a peaceful nuclear cooperation agreement.[23] Section 123 of the Atomic Energy Act (AEA) specifies that proposed nuclear cooperation agreements are to include the terms, conditions, duration, nature, and scope of cooperation. It also requires that any such agreement meet a series of nonproliferation criteria and that the President submit any such agreement to the House Committee on Foreign Affairs and the Senate Committee on Foreign Relations. The Department of State is required to provide the President an unclassified Nuclear Proliferation Assessment Statement (NPAS), which the President is to submit to the committees of referral along with the agreement. The State Department also is required to provide a classified annex to the NPAS, prepared in consultation with the Director of National Intelligence. The NPAS is meant to explain how a proposed agreement would meet the aforementioned nonproliferation criteria. The President also must make a written determination "that the performance of the proposed agreement will promote and will not constitute an unreasonable risk to, the common defense and security." President Bush issued such a determination November 14, 2008.[24] President Obama issued an identical determination May 19, 2009, and submitted the agreement, along with the unclassified NPAS, May 21. President Obama also submitted the classified NPAS.

Under the AEA, Congress has the opportunity to review a 123 agreement for two time periods totaling 90 days of continuous session.[25] The President must submit the text of the proposed nuclear cooperation agreement, along with required supporting documents (including the unclassified NPAS) to the House Foreign Affairs Committee and the Senate Foreign Relations Committee. The President is to consult with the committees "for a period of not less than 30 days of continuous session." After this period of consultation, the President is to submit the agreement to Congress, along with the classified annex to the NPAS and a statement of his approval of the agreement and determination that it will not damage the national security interests of the United States. This action begins the second period, which spans 60 days of continuous session. In practice, the President has submitted the agreement to Congress, along with the unclassified NPAS, its classified annex, and his approval and determination, at the beginning of the full 90- day period. The 60-day period has been considered as following immediately upon the expiration of the 30-day period. If the President has not exempted the agreement from any requirements of Section 123a., it becomes effective at the end of the 60-day period unless, during that time, Congress adopts a joint resolution disapproving the agreement and the resolution becomes law.[26] The proposed agreement with the UAE is not an exempt agreement.

In the 110th Congress, some Members of Congress proposed several amendments to the AEA that would have changed the AEA's procedures for the negotiation and approval of peaceful nuclear cooperation agreements.[27] For example, H.R. 7316, which Representative Ileana Ros-Lehtinen introduced in December 2008, would have required Congress to enact a joint resolution of approval before any peaceful nuclear cooperation agreement could become effective. As noted above, such agreements currently become effective unless Congress enacts a joint resolution of disapproval. The bill also proposed adding a section to the AEA which would have required the President to keep the House Foreign Affairs Committee and the Senate Foreign Relations Committee "fully and currently informed of any initiative or negotiations relating to a new or amended agreement for peaceful nuclear cooperation ...prior to the President's announcement of such initiative or negotiations." The proposed section also would have mandated periodic Presidential consultation with the committees about the progress of negotiations concerning such agreements. In the 111th Congress, H.R. 547, which Representative Ros-Lehtinen introduced January 15, 2009, contains the same language.

ISSUES FOR CONGRESS

Although the final text of the proposed U.S.-UAE nuclear agreement was agreed in early November 2008, the Bush Administration, reportedly at the UAE's request, did not submit the agreement to the 110th Congress.[28] After the Obama Administration took office, the UAE agreed to reopen the text for negotiation. On May 21, 2009, the Administration submitted the agreement to Congress to begin the consultation periods required under the AEA. Some Members of Congress welcomed the UAE government's stated commitments to foreswear proliferation-sensitive nuclear capabilities, such as uranium enrichment or spent fuel reprocessing. Other Members signaled their intention to evaluate the proposed bilateral agreement in light of parallel and specific concerns about the UAE's cooperation with international efforts to prevent Iran from developing nuclear weapons and the potential proliferation or safety risks inherent to exporting U.S. nuclear technology.[29] Broader diplomatic implications of the proposed agreement also were being weighed by concerned parties on all sides. The agreement, however, entered into force following the December 17, 2009 exchange of diplomatic notes because the 90 days of continuous session expired on October 17, 2009. State Department spokesperson Ian Kelly told reporters October 22 that the United States has "completed all ... internal procedures" for the agreement to enter into force. The UAE cabinet approved the agreement October 26.[30]

Congressional Concerns

Export Control Concerns

Since 2001, the UAE has been under increased U.S. scrutiny as an alleged transshipment point for military and dual-use exports to Iran, as an alleged hub of operations for weapons proliferators,[31] and as an alleged transit zone and financial conduit for terrorists and money launderers. At present, particular attention remains focused on U.S. concerns about the UAE government's willingness and ability to halt transfers of militarily sensitive technology to Iran. Since late 2008, some Members of Congress have claimed that the UAE has not acted sufficiently to halt transfers of militarily sensitive technology to Iran and should not be able to conclude a nuclear cooperation agreement with the United States until the UAE government has taken additional measures against Iranian procurement activities of concern. Administration and UAE officials highlight steps taken by the UAE in recent years to strengthen export controls and to take action against entities suspected of illicit proliferation activities, including targets associated with Iran.

The United States government has stated publicly that some UAE-based entities are involved in Iranian weapons-related procurement activities. For example, an October 2008 Department of Justice fact sheet states that eight companies, five of which are based in the

UAE, were charged the previous month with crimes related to their participation in exporting dual-use items to Iran. Additionally, two men were indicted in July 2008 for "participation in a conspiracy to export U.S.-made military aircraft parts to Iran" via a company based in the UAE, according to the fact sheet.[32]

Evidence also suggests that UAE-based entities are involved in procurement activities connected to Iran's nuclear and ballistic missile programs. For example, the Department of the Treasury designated two such entities under Executive Order 13382, which freezes assets under U.S. jurisdiction belonging to designated foreign entities engaged in activities related to the proliferation of Weapons of Mass Destruction (WMD). Specifically, the Treasury Department designated Oriental Oil Kish in October 2007 for unspecified "proliferation activities." Oriental Oil Kish is an entity affiliated with Iran's Islamic Revolutionary Guard Corps (IRGC), a department press release said, adding that the IRGC develops and tests ballistic missiles. In September 2008, the Treasury Department similarly designated Oasis Freight Agencies as a company affiliated with the Islamic Republic of Iran Shipping Lines (IRISL). Adam Szubin, director of the department's Office of Foreign Assets Control, indicated during a September 10, 2008, press conference that IRISL has procured items for Iran's ballistic missile programs.

The Treasury Department has also designated other UAE-based entities under the same executive order because of their ties to Iranian banks which, according to the United States, are involved in proliferation activities. For example, the Department designated Melli Investment Holding International and BMIIC International General Trading Ltd. in March 2009 because of their ties to Iran's Bank Melli, which, according to a March 3 Treasury Department statement, has been involved in procurement activities and other forms of support for Iran's nuclear and missile programs. More recently, the Treasury Department announced May 12, 2009, that it had specifically designated several UAE-based branches of that bank.[33] Additionally, the Department the same day designated the UAE office of Persia International Bank, a subsidiary of Bank Mellat. The Department stated in October 2007 that Bank Mellat has provided "banking services in support of Iran's nuclear entities."

To place UAE-based entities' proliferation role in context, the U.S. government had, as of March 3, 2009, designated approximately 107 non-UAE companies, based in Iran and elsewhere, under Executive Order 13382, though not necessarily for exports to Iran. Moreover, the UAE is not the only conduit for suspicious goods destined for Iran; Tehran has also used a network based in Malaysia for procuring dual-use items.

Concerns about suspicious transfers to Iran prompted U.S. action in 2007 to encourage the UAE to improve its national export control system. In February 2007, the U.S. Department of Commerce released an advanced notice of proposed rule-making that would have created a new export control designation known as "Country Group C" that would have established license requirements on exports and re-exports to countries that represent a diversion or transshipment risk for goods subject to the Export Administration Regulations.[34] Although no countries were mentioned in the notice, the proposal was widely considered to be directed at the UAE.[35]

In August 2007, the UAE adopted a stronger national export control law, but, as of September 2009, the government had yet to issue implementing regulations for the law or to fully staff a national export control body to enforce it. In the interim, export control enforcement functions remain the responsibility of authorities in the UAE's individual emirates, and are being carried out in coordination with a newly established national interagency Committee on Commodities Subject to Import and Export Control. The Committee was authorized in April 2009. The Foreign Ministry serves as the chair, and other members include the Interior Ministry, the Armed Forces, the Ministry of Economy, and the Federal Customs Authority. The Committee met in May 2009 and reportedly discussed "creating a mechanism to implement the law."[36] Minister of State for Foreign Affairs Dr. Anwar Mohammed Gargash said in a statement released in conjunction with the meeting that, "We will not compromise on issues of security and our export control reflects our intention to ensure tough safeguards over the movement of sensitive materials." UAE Ambassador to the United States Yousef Al Otaiba described the law as a "work in progress" during a June 3, 2009, briefing.

U.S. cooperation with national and emirate level officials on proliferation issues appears to be strong. A bilateral nonproliferation working group meets annually to review and discuss nonproliferation issues of shared concern. In a September 2008 letter to then-U.S. Secretary of Commerce Carlos Gutierrez, Ambassador Otaiba detailed six joint and unilateral interdictions on Iran-bound ships completed since June 2008.[37] More recently, the UAE seized "arms and related materials" from a ship en route to Iran, according to Australian officials. The UAE conducted the seizure, which reportedly took place in late July or early August 2009, pursuant to UN Security Council resolution 1874.[38] The UN committee charged with monitoring implementation of the sanctions is investigating the matter, U.S. Permanent Representative to the United Nations Ambassador Susan Rice told reporters September 14.

Otaiba's letter further stated that "the UAE fully supports and has vigorously enforced United Nations resolutions barring the shipment of sensitive materials and technologies to Iran." The UAE also has "closed dozens of international and local companies involved in the transshipment of dual-use and controlled materials," according to the letter, which also highlighted the government's participation in several U.S. security initiatives, including the Container Security Initiative, the Proliferation Security Initiative, and the Department of Energy Megaports Initiative. UAE officials report they remain committed to fully implementing the 2007 law at the national level, including clarifying roles and responsibilities for export control enforcement.

The UAE's Central Bank has also taken steps to pressure Iran. According to the February 2009 International Narcotics Control Strategy Report, the bank "has ordered financial institutions to cease transfers on behalf of designated entities and to refrain from entering into new commitments for grants, financial assistance, and concession loans to the Iranian Government."

The United Kingdom's Export Control Organization has also named two entities located in the UAE, Mileace and Tammam Trading, as possibly being involved in procurement activities for Iran's WMD programs. [39]

Nonproliferation Concerns

The most proliferation-sensitive part of a nuclear power program is the capability to produce fuel for nuclear reactors, either by enriching uranium or reprocessing spent nuclear fuel to obtain plutonium. Low-enriched uranium is used as fuel for nuclear reactors. Both highly enriched uranium and plutonium can be used as fuel in some types of nuclear reactors but are also used as fissile material in nuclear weapons. The dual-use nature of nuclear fuel facilities frequently generates concern that ostensibly peaceful facilities may aid nuclear weapons programs.

The April MOU states that the UAE has agreed to the policy commitments described earlier in its April 2008 policy statement, which are designed to boost confidence that the state's nuclear program is exclusively for peaceful purposes. For example, the statement indicates that the UAE will forgo "domestic enrichment and reprocessing capabilities in favor of long-term commitments of the secure external supply of nuclear fuel." Moreover, as noted above, the nuclear cooperation agreement's text states that the United States can end nuclear cooperation with the UAE if it acquires enrichment or reprocessing facilities. Without such capabilities, a nuclear program poses little proliferation risk. IAEA Director-General Mohamed ElBaradei explained in an August 2007 interview:

> One nuclear reactor by itself means nothing, you are still far from having an atom bomb. I am more worried when a country has a plant for industrial-scale uranium enrichment...In this case it can make a nuclear bomb within a few months. [40]

As noted above, the U.S. State Department reports that the UAE's new nuclear regulatory law (Federal Law 6 of 2009) prohibits domestic enrichment and reprocessing.[41]

As a party to the nuclear Nonproliferation Treaty (NPT), any future UAE nuclear facilities would be subject to IAEA safeguards.[42] Additionally, the UAE agreed to conclude an Additional Protocol to its safeguards agreement. As noted, such protocols give IAEA officials greater access to an NPT state's nuclear-related facilities and information. The UAE currently has a Small Quantities Protocol to its safeguards agreement, but, according to the nuclear cooperation agreement, will terminate that Protocol before the United States issues export licenses for the export of "nuclear material, equipment, components, or technology" pursuant to the cooperation agreement.[43]

It is also worth noting that the UAE's 2008 policy statement on its nuclear program states that the government plans to rely on light-water reactors, which are considered among the most proliferation-resistant, partly because of the difficultly in producing and obtaining weapons-grade plutonium without detection. Moreover, a May 2008 International Institute for Strategic Studies report points out that "no successful nuclear-weapons program has ever relied on commercial reactors."[44] Although a civilian nuclear power program could provide cover for a country's procurement of dual-use items that could aid a nuclear weapons program, such a program would need to include some covert facilities.

Human Rights Concerns

A video depicting the torture of an Afghan grain merchant named Muhammad Shah Poor by Abu Dhabi ruling family member Sheikh Issa bin Zayed al Nahayan and uniformed security officers has drawn widespread condemnation following its publication by ABC News in April 2009.[45] In response, some Members of Congress and congressional staff have raised questions about the appropriateness of moving forward with the proposed U.S.-UAE nuclear cooperation agreement, pending UAE action on the case.[46] The U.S. State Department has indicated that the Obama Administration's review of the proposed nuclear cooperation agreement and concerns the Administration may have about the torture depicted on the video "are two separate issues."[47] Abu Dhabi authorities announced on May 11, 2009, that "the Public Prosecution Office has officially launched a criminal investigation into the events depicted on video, and detained Sheikh Issa bin Zayed Al Nahyan pending the outcome of this investigation."[48]

Possible Diplomatic Implications

Shaping Nonproliferation Standards and Best Practices

The Bush Administration argued that nuclear cooperation with the UAE could set a useful precedent for mitigating the dangers of nuclear proliferation as an increasing number of countries consider developing nuclear power. The State Department stated in April 2008 that the UAE's choice to forgo enrichment and reprocessing "serves as a model for the economical and responsible pursuit of nuclear power." Similarly, President Obama's May 21 letter of transmittal argued that the agreement "has the potential to serve as a model for other countries in the region that wish to pursue responsible nuclear energy development." As noted, the proposed U.S.-UAE agreement includes a provision which apparently intends to establish the agreement's conditions as minimum standards for future such agreements in the Middle East. On July 8, 2009, Under Secretary of State for Arms Control and International Security Ellen Tauscher testified before the House Foreign Affairs Committee on the proposed agreement and argued that

> In addition to being indicative of our strong partnership with the UAE, the proposed Agreement is a tangible expression of the United States' desire to cooperate with states in the Middle East, and elsewhere, that want to develop peaceful nuclear power in a manner consistent with the highest nonproliferation, safety, and security standards… U.S. cooperation with the UAE will also serve as a distinct counterpoint to those countries that have chosen a different path, in particular Iran.[49]

Commercial Opportunities

Licensed nuclear technology contracts with the UAE could provide commercial benefits to the U.S. nuclear industry or its international competitors. While Emirati officials have stated their strong desire for nuclear cooperation with the United States and have incorporated former U.S. government officials and U.S. contractors into their early plans and activities, the UAE also is seeking similar cooperation from other international sources. The expected value of eventual construction and management contracts is in the tens of billions of dollars. The UAE and France signed a nuclear cooperation agreement in with France in January 2008 and a similar agreement with South Korea June 22, 2009. The UAE has also signed memoranda of understanding with Japan and the United Kingdom concerning nuclear technology.

UAE officials explained during a June 3, 2009, briefing that the government's nuclear cooperation agreement with France contains preambulary language describing the UAE's commitment to refrain from enrichment and reprocessing. However, the agreement does not explicitly refer to that commitment. Transferring such technologies is not within the agreement's scope, the officials said, adding that all future UAE nuclear cooperation agreements with other countries will reflect the government's policy regarding enrichment and reprocessing.

Bilateral Relations and UAE Cooperation in Nonproliferation Activities

As noted above, the United States and the UAE have enjoyed close and cooperative relations in recent years, in spite of periodic differences over some issues. Some observers have speculated that a failure to conclude the proposed nuclear cooperation agreement would be viewed by officials and influential figures in the UAE as an indication of a lack of faith and commitment by the United States government in the UAE, which could have negative implications for other aspects of the bilateral relationship. These fears appear to be based largely on the perceived repercussions of Dubai Ports World's failed 2006 bid to acquire and operate U.S. port terminals. However, others believe that while the Dubai Ports World incident undermined feelings of mutual trust and set back some planned commercial ties, the failed initiative did not otherwise damage U.S.-Emirati political or military relations in any tangible, lasting way.

By all accounts, Emirati authorities appear committed to moving forward with their nuclear development plans and, while they appear to strongly desire U.S. technical assistance, it seems unlikely that their plans would be curtailed or significantly undermined if the United States declined to authorize the proposed agreement. As noted above, other international parties appear willing to provide technical assistance on a commercial basis. Some observers have argued that without U.S. involvement, the UAE program could adopt technology or systems more vulnerable to proliferation or security concerns. UAE representatives state that their commitment to forego domestic enrichment or reprocessing is fundamental and would apply under cooperative arrangements with non-U.S. suppliers.

Legislative Developments in the 110th and 111th Congress

In the 110th Congress, Representative Ros-Lehtinen introduced H.R. 7316, the "Limitation on Nuclear Cooperation with the United Arab Emirates Act of 2008." The bill would have prohibited the proposed U.S.-UAE agreement from coming into effect without presidential certification that the UAE had taken specific steps to improve its export controls and to limit the transfer of certain items to Iran. As noted above, H.R. 7316 also sought to change the procedures for the negotiation and approval of peaceful nuclear cooperation agreements by amending the Atomic Energy Act. (See "Nuclear Cooperation Agreements, Approval Process, and Proposed Changes" above.)

In the 111th Congress, Representative Markey introduced a joint resolution of disapproval on May 21, 2009 (H.J.Res. 55). House and Senate joint resolutions of approval were introduced on July 13, 2009 (S.J.Res. 18 and H.J.Res. 60). Representative Ros-Lehtinen is an original co-sponsor of H.J.Res. 60, although she had previously introduced two pieces of legislation to establish conditions on U.S.-UAE nuclear cooperation and to change negotiation and approval procedures for nuclear cooperation agreements:

- H.R. 364, the "Limitation on Nuclear Cooperation with the United Arab Emirates Act of 2009" was introduced and referred to the House Foreign Affairs Committee on January 9, 2009. The bill states that a U.S. civil nuclear cooperation agreement with the UAE may not enter into force "unless not less than 30 legislative days prior to such entry into force the President certifies" to the House Foreign Affairs Committee and the Senate Foreign Relations Committee that the UAE has improved its export control system and halted UAE-based entities' transfers of technology relating to Weapons of Mass Destruction programs, particularly to Iran. The bill also requires the UAE to stop the transfer of certain conventional weapons and related components to Iran. H.R. 364 also states that, if the United States and the UAE do conclude a nuclear cooperation agreement, the United States may not grant an export license for "nuclear material, equipment, or technology" to the UAE unless the President certifies within 30 legislative days that the UAE has met the above requirements.

- H.R. 547, was introduced and referred to the House Foreign Affairs Committee January 15, 2009. It contains the language from H.R. 7316 relating to procedural changes for nuclear cooperation agreements pursuant to the Atomic Energy Act.

Appendix. Provisions in U.S. Nuclear Cooperation Agreements with the UAE and Egypt Relevant to Establishing Standards for other Such Agreements

UAE

The Government of the United States of America confirms that the fields of cooperation, terms and conditions accorded by the United States of America to the United Arab Emirates for cooperation in the peaceful uses of nuclear energy shall be no less favorable in scope and effect than those which may be accorded, from time to time, to any other non-nuclear-weapon State in the Middle East in a peaceful nuclear cooperation agreement. If this is, at any time, not the case, at the request of the Government of the United Arab Emirates the Government of the United States of America will provide full details of the improved terms agreed with another nonnuclear-weapon State in the Middle East, to the extent consistent with its national legislation and regulations and any relevant agreements with such other non-nuclear weapon State, and if requested by the Government of the United Arab Emirates, will consult with the Government of the United Arab Emirates regarding the possibility of amending this Agreement so that the position described above is restored.

EGYPT

The Government of the United States confirms that fields of cooperation, terms and conditions accorded by the United States to the Arab Republic of Egypt for cooperation in the peaceful uses of nuclear energy shall be no less favorable in scope and effect than those which may be accorded by the United States to any other non-nuclear weapon state in the Middle East in a peaceful nuclear cooperation agreement. In this connection it is understood that the safeguards required by this agreement shall be no more restrictive than those which may be required in any peaceful nuclear cooperation agreement between the United States and any other state in the region. By entering into this agreement the United States confirms its recognition of the importance of the Arab Republic of Egypt's adherence to the NPT, and its longstanding support of international non-proliferation measures, including establishment of a nuclear weapon free zone in the Middle East. If any situation arises which could increase the risk of proliferation of nuclear weapons, the United States and the Arab Republic of

Egypt, at the request of either, shall enter into consultations with respect thereto with a view to maintaining the objectives of the NPT.

End Notes

[1] The governments of Turkey, Egypt, Jordan, Saudi Arabia, and Algeria have announced their intent to acquire nuclear energy production capabilities; their respective programs have moved forward in recent years with varying degrees of specificity and commitment. To date, Turkey, Jordan, and Egypt appear to have made the most progress toward their stated goals of constructing and operating domestic nuclear power plants. Like the UAE, their plans do not envision operational plants before 2015. Turkey and Egypt have active peaceful nuclear cooperation agreements with the United States pursuant to Section 123 of the Atomic Energy Act (AEA) of 1954.

[2] Policy of the United Arab Emirates on the Evaluation and Potential Development of Peaceful Nuclear Energy, released April 20, 2008. Available at https://pcs.enec.gov.ae/Content/Home.aspx.

[3] For more detail about the proposal, see CRS Report RL34234, *Managing the Nuclear Fuel Cycle: Policy Implications of Expanding Global Access to Nuclear Power*, coordinated by Mary Beth Nikitin.

[4] CRS meeting with Abu Dhabi Executive Authority and Emirates Nuclear Energy Corporation representatives, Abu Dhabi, December 14, 2008. Large scale petrochemicals and aluminum production projects are already underway in Abu Dhabi's Kalifa Industrial Zone. See Chris Stanton, "Taweelah to host chemical city," *The National* (Abu Dhabi), May 4, 2008; and, Chris Stanton and Ivan Gale, "EMAL smelter remains on schedule," *The National* (Abu Dhabi), January 13, 2009.

[5] "Policy of the United Arab Emirates on the Evaluation and Potential Development of Peaceful Nuclear Energy," released April 20, 2008. Available at https://pcs.enec.gov.ae/Content/Home.aspx.

[6] One economic press report suggested that the UAE may be paying as little as $1.25/million BTU of natural gas. *Middle East Economic Digest*, "UAE purchases gas from Dolphin pipeline at reduced rate," May 16, 2008. For more information on the pipeline, see http://www.oxy.com/Our_Businesses/oil_and_gas/Pages/og_mena_dolphin.aspx.

[7] Correspondence between CH2M Hill and U.S. Department of Energy, National Nuclear Security Administration, June 2008.

[8] CRS meeting with Abu Dhabi Executive Authority and Emirates Nuclear Energy Corporation representatives, Abu Dhabi, December 14, 2008.

[9] See IAEA-Technical Cooperation, National and Regional Asia and the Pacific Projects query page at http://www-tc.iaea.org/tcweb/tcprogramme/recipients/ eastasiapacific/ query/default.asp.

[10] Statement of Vann H. Van Diepen, Acting Assistant Secretary, State Department Bureau of International Security and Nonproliferation before the Senate Foreign Relations Subcommittee on Near Eastern and South and Central Asian Affairs, October 7, 2009.

[11] CRS analyst interview with UAE official, Washington, DC, March 6, 2009.

[12] UAE Minister of Foreign Affairs Shaykh Abdullah bin Zayed Al Nahyan, "Why Go Nuclear?" *Bulletin of the Atomic Scientists*, September/October 2008; and, CRS meeting with Abu Dhabi Executive Authority and Emirates Nuclear Energy Corporation representatives, Abu Dhabi, December 14, 2008.

[13] "Policy of the United Arab Emirates." The four instruments are: the Vienna Convention on Civil Liability for Nuclear Damage, the Protocol to Amend the Vienna Convention on Civil Liability, the Joint Protocol on the Application of the Vienna and Paris Conventions, and the Convention on Supplementary Compensation for Nuclear Damage.

[14] See ENEC homepage at https://pcs.enec.gov.ae/Content/Home.aspx.

[15] For more information on the Partnership, see CRS Report RL34234, *Managing the Nuclear Fuel Cycle: Policy Implications of Expanding Global Access to Nuclear Power*, coordinated by Mary Beth Nikitin.

[16] Uranium enrichment and reprocessing spent nuclear fuel are the dual-use nuclear activities of greatest proliferation concern. Uranium enrichment can produce low-enriched uranium for nuclear reactor fuel or highly enriched uranium, which can be used as fissile material in nuclear weapons. Reprocessing spent nuclear fuel separates plutonium from that fuel. Plutonium can also be used as fissile material in nuclear weapons.

[17] According to the agreement text, "special fissionable material means (1) plutonium, uranium 233, or uranium enriched in the isotope 235, or (2) any other material so designated by agreement of the Parties." A similar provision is contained in a 1981 U.S. nuclear cooperation agreement with Egypt.

[18] CRS Analyst interview with State Department official March 9, 2009.

[19] U.S. agreements with Finland and Sweden also granted such consent rights, but those agreements have been replaced by the 1996 U.S.-EURATOM agreement.

[20] Japan and Switzerland have received advance consent to take back recovered plutonium for use in their civil nuclear programs.

[21] CRS Analyst interviews with State Department official March 12, 2009 and former State Department official Fred McGoldrick March 10, 2009.

[22] Significant nuclear cooperation includes the transfer of U.S.-origin special nuclear material subject to licensing for commercial, medical, and industrial purposes. The term "special nuclear material," as well as other terms used in the statute, is defined in 42 U.S.C. §2014. "Special nuclear material" means (1) plutonium, uranium enriched in the isotopes 233 or 235, and any other material that is determined to be special nuclear material, but does not include source material, or (2) any material artificially enriched by any of the foregoing, but does not include source material.

[23] For a primer on such agreements, which are frequently referred to as "123 agreements," see CRS Report RS22937, *Nuclear Cooperation with Other Countries: A Primer*, by Paul K. Kerr and Mary Beth Nikitin.

[24] Presidential Determination No. 2009–7, November 14, 2008.

[25] When calculating periods of "continuous session" under the AEA, every calendar day is counted, including Saturdays and Sundays. Only days on which either chamber has adjourned for more than three days pursuant to the adoption of a concurrent resolution authorizing the adjournment do not count toward the total. If Congress adjourns its final session *sine die*, continuity of session is broken, and the count must start anew when it reconvenes.

[26] The AEA allows the President to exempt a proposed nuclear cooperation agreement from the nonproliferation criteria specified in Section 123. Such agreements have different procedures for Congressional review. Since the proposed agreement with the UAE is apparently a non-exempt agreement, those procedures are not discussed here.

[27] For additional details on these proposals, see CRS Report RS22937, *Nuclear Cooperation with Other Countries: A Primer*, by Paul K. Kerr and Mary Beth Nikitin.

[28] "UAE, USA Complete Negotiations on Peaceful Nuclear Energy Agreement," Emirates News Agency, December 15, 2008; and, Mark Hibbs, "US-UAE Cooperation Pact Initialed; Approval Left to Obama, New Congress," Nucleonics Week, November 20, 2008.

[29] See, for example, a November 24, 2008, statement from Representative Brad Sherman, a January 15, 2009, letter from Representative Edward Markey, and an April 6, 2009, letter from Markey, Sherman, and Representative Ileana Ros-Lehtinen.

[30] "Cabinet Approves UAE-U.S. Agreement on Peaceful Nuclear Cooperation," *Emirates News Agency*, October 26, 2009.

[31] In connection with revelations of illicit sales of nuclear technology to Iran, Libya, and North Korea by Pakistan's nuclear scientist A.Q. Khan, Dubai was named as a key transfer point for Khan's shipments of nuclear components. For more detail, see CRS Report RL32745, *Pakistan's Nuclear Proliferation Activities and the Recommendations of the 9/11 Commission: U.S. Policy Constraints and Options*.

[32] "Fact Sheet: Major U.S. Export Enforcement Prosecutions During Past Two Years," U.S. Department of Justice, October 28, 2008.

[33] Previously, all worldwide offices of Bank Melli had been designated, but they were not named specifically.

[34] "Country Group C: Destinations of Diversion Control," Advanced Notice of Proposed Rulemaking, 72 Federal Register 8315, February 26, 2007.

[35] *Inside U.S. Trade*, "UAE Drafts New Export Control Law With U.S. Help," Vol. 25, No. 11, March 16, 2007.

[36] Mahmoud Habboush, "Arms export control committee meets," *The National* (Abu Dhabi), May 10, 2009.

[37] Letter from UAE Ambassador to the United States Yousef Al Otaiba to U.S. Secretary of Commerce Carlos Guitierrez, September 17, 2008.

[38] Brendan Nicholson, "Australian Ship in Arms Row," *The Age*, August 31, 2009; Rod Mcguirk, "Australia Probes North Korean Weapons Shipment," The Associated Press, August 30, 2009. For additional information on the Security Council resolution, see CRS Report R40684, *North Korea's Second Nuclear Test: Implications of U.N. Security Council Resolution 1874*, coordinated by Mary Beth Nikitin and Mark E. Manyin.

[39] "Iran List - WMD End-Use Control: Licence Applications for Iran," updated August 28, 2009. Available at http://www.berr.gov.uk/whatwedo/europeandtrade/strategic-export-control/licensing-policy/end-use-control/page29307.html.

[40] "We're approaching the brink: PROFIL Interview with Mohamed El Baradei." *Profil*, August 8, 2007. Available at http://www.iaea.org/NewsCenter/Transcripts/2007/PROFIL3507.html.

[41] Statement of Vann H. Van Diepen, Acting Assistant Secretary, State Department Bureau of International Security and Nonproliferation before the Senate Foreign Relations Subcommittee on Near Eastern and South and Central Asian Affairs, October 7, 2009.

[42] For more information about IAEA safeguards, see CRS Report RL33865, *Arms Control and Nonproliferation: A Catalog of Treaties and Agreements*, by Amy F. Woolf, Mary Beth Nikitin, and Paul K. Kerr. The UAE has had an IAEA safeguards agreement in force since 2003. The text is available at http://www.iaea.org/Publications/Documents/ Infcircs/2003/infcirc622.pdf.

[43] Some NPT state-parties with small quantities of fissionable materials have concluded a Small Quantities Protocol to their IAEA safeguards agreements. Certain IAEA verification requirements are suspended for such states.

[44] International Institute for Strategic Studies, *Nuclear Programmes in the Middle East: In the Shadow of Iran*, May 2008.

[45] Vic Walter, Rehab El-Buri, Angela Hill, and Brian Ross, "Torture Tape Implicates UAE Royal Sheikh," *ABC News: Nightline*, April 22, 2009. Available at http://abcnews.go.com/Blotter/Story?id=7402099&page=1.

[46] For example, Congressman James McGovern, the co-chairman of the House Human Rights Commission has asked, "How can we move forward with such a delicate agreement in the face of such an atrocious human rights violation?" adding, "If it [the proposed 123 agreement] were brought to Congress now I would certainly ask that it be rejected." Robert F. Worth, "Abu Dhabi Torture Video Raises Doubts in U.S. About Nuclear Pact," *New York Times*, May 2, 2009.

[47] Robert Wood, Acting State Department Spokesman, Daily Press Briefing, Washington, DC, April 30, 2009.

[48] *Emirates News Agency* (WAM), "Video matter referred to the Abu Dhabi Public Prosecution Office," May 11, 2009.

[49] Testimony of Under Secretary Ellen Tauscher Before the House Foreign Affairs Committee, July 8, 2009.

CHAPTER SOURCES

Chapter 1 - This is an edited, reformatted and augmented version of a Congressional Research Service publication, RS22937, dated December 8, 2010.

Chapter 2 - This is an edited, reformatted and augmented version of a U.S. Government Accountability Office publication, GAO-11-36, dated November, 2010.

Chapter 3 - This is an edited, reformatted and augmented version of a Congressional Research Service publication, R41312, dated December 1, 2010.

Chapter 4 - This is an edited, reformatted and augmented version of a Congressional Research Service publication, RL34655, dated September 17, 2010.

Chapter 5 - This is an edited, reformatted and augmented version of a Congressional Research Service publication, RL33016, dated October 28, 2010.

Chapter 6 - This is an edited, reformatted and augmented version of a Congressional Research Service publication, R41032, dated January 21, 2010.

Chapter 7 - This is an edited, reformatted and augmented version of a Congressional Research Service publication, R40344, dated December 23, 2009.

INDEX

O

P

Q

R